OFF-SITE ENHANCED BIOGAS PRODUCTION WITH CONCOMITANT PATHOGEN REMOVAL FROM FAECAL MATTER

T0136392

Joy Nyawira Riungu

OFF-SITE ENHANCED BIOGAS PRODUCTION WITH CONCOMITANT PATHOGEN REMOVAL FROM FAECAL MATTER

DISSERTATION
Submitted in fulfilment of the requirements of
the Board for Doctorates of Delft University of Technology
and
of the Academic Board of the IHE Delft
Institute for Water Education
for
the Degree of DOCTOR
to be defended in public on
Thursday, 28 January 2021, at 15:00 hours
in Delft, the Netherlands

by

Joy Nyawira RIUNGU
Master of Science in Environmental Engineering and Management, Jomo Kenyatta
University of Agriculture and Technology, Nairobi, Kenya
Born in Meru, Kenya

This dissertation has been approved by the
promotor: Prof.dr.ir. J.B. van Lier
copromotor: Dr.ir. M. Ronteltap

Composition of the doctoral committee:

Rector Magnificus TU Delft	Chairman
Rector IHE Delft	Vice-Chairman
Prof.dr.ir. J.B. van Lier	IHE Delft / TU Delft, promotor
Dr.ir. M. Ronteltap	IHE Delft, copromotor

Independent members:
Prof.dr.ir. M.K. de Kreuk	TU Delft
Prof.dr.ir. G. Zeeman	Wageningen University & Research
Prof.dr. C. Chernicharo	Universidade Federal de Minas Gerais, Brazil
Prof.dr. F. Kansiime	Makarere University, Uganda
Prof.dr.ir. L.C. Rietveld	TU Delft, reserve member

This research was conducted under the auspices of the Graduate School for Socio-Economic and Natural Sciences of the Environment (SENSE)

CRC Press/Balkema is an imprint of the Taylor & Francis Group, an informa business

Published by:
CRC Press/Balkema
Schipholweg 107C, 2316 XC, Leiden, the Netherlands
Pub.NL@taylorandfrancis.com
www.crcpress.com – www.taylorandfrancis.com
ISBN: 978-1-032-00443-3 (Taylor and Francis Group)

Dedication

This thesis is dedicated to the research community interested in resource-oriented sanitation.

Acknowledgement

The greatest motivation of starting my PhD research was to gain knowledge adequate to make contribution to faecal sludge management research. This was made possible through financial help by Bill & Melinda Gates Foundation under the framework of SaniUp project (Stimulating local Innovation on Sanitation for the Urban Poor in Sub-Saharan Africa and South-East Asia) under leadership of Prof. Damir Brdjanovic. I am also grateful to Government of Kenya, and Meru University of Science and Technology for all the support that I received in the course of my research.

In realising my dream, the role played by other people has been incredible; words may not adequately express! My Promotor, Professor Jules J.B. van Lier and my mentor Dr. Mariska Ronteltap; have learnt so much more from your supervision and mentorship than my PhD thesis can convey. Prof. Jules, you encouraged me and made me believe in myself whenever things were not working. Your guidance and supervision style is amazing. Dr Mariska, your dedication to my PhD supervision, and the way you combined it so well with your other commitments will always be a driving force for me. I have received massive support and goodwill from colleagues just by being your supervisee.

My UNESCO shit team: Laurens Welles, Fiona Zakaria, Peter Mawioo and Xavier Sanchez, Sondos Saad, the office fun times made life bearable despite the pressure. Christine Etiegni, my weekends in Delft would not have been complete without you, thanks very much. My spiritual parents at Delft, Paster Waltrout and Fr Peter, thank you so much for providing spiritual nourishment. Ani Vallabhaneni and David Auerbach, Sanergy Kenya, my laboratory work would not have been complete without the immense support received from you and your staff. In particular, special regards to James Kaburu, Edwin Wekesa, Lisbeth Wangari and Naomi for support with field work; sample collection and analysis.

I am also very thankful to my family for being my friends, dependable support, and loudest cheerleaders, particularly son Mark Murimi and Clement Munene, my parents mum Speranza Kagoji and late father Justus Riungu, and my miraculous brothers and sisters. Boniface Mutunga, thanks for being my confidant; literally your listening ear has made life manageable for me.

Thesis summary

Effective and sustainable faecal sludge management (FSM) poses a sanitation challenge in developing countries. It is exacerbated by high population growth rate, emergence of low income, high density settlements (LIHDS) and social cultural perceptions surrounding human faecal sludge (FS) handling. FS handling has been relegated to a passive exercise whose end point is to get the waste from the immediate vicinity with no follow-up on its subsequent endpoint. While the stigma associated with human waste is borne from the potential health risk, not handling it poses a definite health hazard to the population.

Previously applied sewer-based sanitation systems are expensive to develop in poor and less developed countries, in addition to significant costs required for maintaining and upgrading infrastructure. Onsite sanitation technologies (in particular pit latrines), previously viewed as sanitation solution for rural areas has widely been adopted in urban cities of developing countries. These however focused on FS collection neglecting emptying, transportation, treatment and disposal/reuse of the end products. As such, they are characterised by a fill-and-abandon strategy with their sustainability limited by availability of land - especially with increasing population. Moreover, while abandoned pit latrines leave the area unusable for other gainful activities, the sludge thus buried finds its way into shallow wells and groundwater as a pollutant and the otherwise rich organic material becomes unavailable for biogeochemical cycling. As such, sustainable and cost-effective faecal sludge management interventions, applicable in informal urban and peri-urban settlements are essential. The interventions should primarily focus on the whole sanitation chain: collection, emptying, transportation, treatment and final disposal/end use.

Resource oriented based approach to sanitation provision is slowly being adopted especially in LIHDS. This, in addition to enhancing sanitation ensures that nutrients available in FS are utilised for betterment of human life e.g. biochemical energy recovery, organic fertiliser, briquette, protein production etc. The approach in Kenya has been adopted and practically used by Sanergy, a social enterprise working on sanitation improvement in LIHDS. They apply use of Urine Diverting Dehydrating Toilets (UDDT) in informal settlements in Nairobi, registering a daily collection of 8 tonnes of faecal matter from UDDTs (UDDT-FS). The UDDT principle involves separate collection of faeces and urine. After use, approximately 15 grams of sawdust are added for odour and fly elimination. After collection, UDDT-FS is transported to a central treatment facility, located at Kinanie, 40 kilometres from Nairobi. The waste undergoes additional treatment steps since bare addition of sawdust does not provide adequate pathogen inactivation. Sanergy, applying composting as the main treatment method, is overwhelmed by the huge amount of waste delivered per day, thus sought to diversify treatment options.

Whereas Sanergy's sanitation model is working, previous attempts in enhancing sanitation in LIHDS failed. This study evaluates failed attempts in LIHDS sanitation enhancement and provides alternative technology for treatment of FS, enabling resource recovery and pathogen inactivation. The study was conducted in 2 phases:
a. Phase 1 – Field work on the limitations of LIHDS sanitation chain; case study Kibera

b. Phase 2i&2ii – Experimental work on biochemical energy recovery and pathogen inactivation during anaerobic stabilisation of UDDT-FS.

a) Phase 1 – Field work

Field work evaluated the limitations of LIHDS sanitation chain. Results are presented from questionnaires and qualitative interviews with stakeholders involved at all levels of sanitation chain i.e. collection, emptying, transportation, disposal, treatment and final disposal/reuse. The provision of sanitation facilities is not adequate in addressing the sanitation challenge in LIHDS. Fill and abandon strategy characterises use of pit latrines within the settlement.

Currently, there is no policy framework governing planning, implementation and management of onsite sanitation at City level. This leads to uncoordinated among the various stakeholders along the sanitation chain. Sanitation providers focus on provision of sanitation facilities, neglecting emptying, transportation, treatment and disposal/reuse of end product. Pit emptiers likewise have unregulated operations with 85%faecal sludge (FS) ending up untreated back to environment. Pay-and-use approach of sanitation provision enhances operation and maintenance initiatives: whereas 73% of free to use facilities were abandoned on fill up, 89% and 77% of community-based organisations (CBOs) and entrepreneur-managed facilities respectively were well managed. Partnership-based sanitation provision improvements provide an entry point for broader initiatives to improve living conditions in informal settlements. By involving government actors, CBOs, community, pit emptiers, Umande Trust and Sanergy have created an economically viable approach of inclusive sustainable sanitation for underprivileged population. Results from the study are useful to the local government and other partners involved in sanitation improvement within the LIHDS.

b) Phase 2i&2ii – Laboratory scale and pilot scale experiments

Alternative FS treatment technologies are needed for management of FS from onsite technologies. Phase 2i and 2ii of study explored potential for biochemical energy recover and pathogen inactivation during anaerobic stabilisation of UDDT-FS at i) Laboratory scale and ii) Pilot scale experiments.

2i. Laboratory scale tests

This study examined the degree of inactivation of *Escherichia coli (E. coli)* and *Ascaris lumbricoides (A. lumbricoides)* eggs in faecal matter obtained from urine diverting dehydrating toilets (UDDT-F) by applying high concentrations of volatile fatty acids (VFAs) during anaerobic digestion. The impact of individual VFAs on *E. coli* and *A. lumbricoides* eggs inactivation in UDDT-FS was assessed by applying various concentrations of store-bought VFA (acetate, propionate and butyrate). High VFA concentrations were also obtained by performing co-digestion of UDDT-FS with organic market waste (OMW) using various mixing ratios. Study findings showed a positive correlation between *E. coli* log inactivation and VFA concentration. For co-digestion, the OMW fraction in the feed substrate was observed to affect VFA build-up, increasing at higher OMW fractions. However, the application of too high OMW fractions is disadvantageous due to: 1) logistic concerns - collection, sorting and transportation costs of the waste from the LIHDS to the treatment site -, and 2) at high OMW fraction (e.g. UDDT-FS:OMW ratios 1:4, 1:2 and 1:1) pH declines to very low levels, which is toxic to the microbial population essential for anaerobic digestion. As such, ND-VFA build-up in the range of 1.2-1.8 meq/g total TS added seems to be sufficient, which agrees with a ND-VFA concentration of approximately 2800-4300 mg/L. In 4 days, between 3 to 5 $E.$ $coli$ log inactivation was achieved at a UDDT-FS:OMW ratio of 2:1 to 4:1. As such, a UDDT-FS:OMW ratio of 4:1 was recommended for further evaluation under pilot scale study. Further, an assessment of biogas production at the recommended UDDT-FS:OMW ratio 4:1 was carried out, where two more control experiments at UDDT-FS:OMW ratio 1:0 and 0:1 were also set. The findings showed higher biogas production at a higher OMW fraction, attributed to a higher hydrolysis noted in previous experiment. However, under practical application, sole digestion of OMW can lead to reactor acidification.

2ii. Pilot scale experiments

Under pilot scale experiments, digestion and co-digestion of faecal matter collected from urine diverting dehydrating toilet faces (UDDT-FS) and mixed organic market waste (OMW) was studied in a single stage and in a two-stage pilot scale mesophilic plug-flow anaerobic reactor at UDDT-FS:OMW ratio 4:1 recommended by laboratory scale study findings. A control experiment at UDDT-FS:OMW ratio 1:0 was set. $E.$ $coli$ inactivation, volatile fatty acids (VFA) build-up and biogas production were monitored at sampling points located along the reactor profile. Total solid (TS) concentration was based on the maximum that could flow without the need for pumping, thus 12% TS applied. 10% TS was applied to assess the effect of pathogen inactivation at a lower TS concentration.

Co-digesting UDDT-FS:OMW in a two stage reactor depicted 8.0 log pathogen inactivation higher than single- stage reactor with 5.7 log inactivation. Highest VFA concentration was 6.3±1.3 g/L, obtained at a pH of 4.9 in the hydrolysis/acidogenesis reactor under the application of a UDDT-FS:OMW ratio of 4:1 and 12% TS, corresponding to a non-dissociated (ND)-VFA concentration of 6.9±2.0 g/L and decay rate of 1.6 /d. In the subsequent methanogenic plug-flow reactor, a decay rate of 1.1/d was attained within the first third part of the reactor length, declining to 0.6/d within the last third part of the reactor length.

In subsequent pilot scale experiments, single- stage and two- stage plug flow digesters were evaluated, applying UDDT-FS: OMW ratio of 4:1 and 1:0, at 10% and 12% TS concentrations. Comparable methane production was observed in single- stage ($R_{s-4:1, 12\%}$) (314±15 mL CH_4 /g VS added) and two- stage ($R_{am-4:1, 12\%}$) (325±12 mL CH_4 /g VS added) digesters, when applying 12% total solids (TS) slurry concentration. However, biogas production in $R_{am-4:1, 12\%}$ digester (571±25 mL CH_4/g VS added) was about 12% higher than in the $R_{s-4:1, 12\%}$. This was attributed to enhanced waste solubilisation and increased CO_2 dissolution, resulting from mixing the bicarbonate-rich methanogenic effluent for neutralisation purposes with the low pH (4.9) influent coming from the pre-acidification stage. Higher process stability was observed in the first parts of the plug flow two- stage digester, characterised by lower VFA concentrations.

The research findings show potential for a two-stage plug-flow reactor as a cost-effective FS treatment technology. For practical purposes, study recommends operating the reactor at UDDT-FS:OMW ratio of 4:1, at 12% TS concentration, 35^0C, thus maximising on the UDDT-FS fraction that could be treated per unit time. A TS concentration of 12% was the highest possible concentration that could flow without the need for mechanical pumping. Increased pathogen inactivation was observed at an increasing OMW fraction in feed substrate but may lead to process failure due to very low pH levels attained, which may be toxic to *E. coli,* as well as all other microbial populations. The system does not only promote hygienic handling of faecal matter but also enables recovery of resources. Biogas produced from the facilities is sold to the LIHDS dwellers, as cooking fuel within the centres or to provide hot showers, with charges depending on the type of service offered and range from 0.2 euro (rice cooking or hot shower) to 0.4 euro (cooking beans). In addition, the effluent can be applied as an alternative fertiliser source to increase agricultural production, replacing commercial fertiliser sources.

Moreover, the system has potential for application as an off-site FS treatment technology at any scale, thus applicable for FS treatment in LIHDS in sub-Saharan Africa. Under off-site arrangement, waste is generated elsewhere and delivered for processing at a central treatment site. On this basis, the technology can be applied to treat faecal waste collected using other technologies e.g. UDDT, peepoo bags etc. To reduce logistics and operation cost of transporting UDDT-FS, off-site treatment sites should be located as near as possible to the UDDT-FS collection points.

Thesis samenvatting

Effectief en duurzaam fecaal slibbeheer (FSM) vormt een uitdaging voor sanitatie in ontwikkelingslanden. De uitdaging wordt versterkt door de hoge bevolkingsgroei, de opkomst van informele sloppenwijken en sociaal-culturele percepties rond de behandeling van menselijk fecaal slib (FS). FS-behandeling is gedegradeerd tot een passieve oefening waarvan het eindpunt is om het afval uit de directe omgeving te halen zonder follow-up op het daaropvolgende eindpunt. Hoewel het stigma in verband met menselijk afval wordt veroorzaakt door het potentiële gezondheidsrisico, vormt het niet-hanteren ervan een duidelijk gezondheidsrisico voor de bevolking.

Elders toegepaste rioleringssystemen zijn duur om te ontwikkelen in arme en minder ontwikkelde landen, naast de aanzienlijke kosten die nodig zijn voor het onderhoud en de verbetering van de infrastructuur. Sanitaire technologieën ter plaatse (met name latrines), die voorheen werden beschouwd als sanitaire oplossing voor landelijke gebieden, worden algemeen toegepast in stedelijke gebieden van ontwikkelingslanden. Omdat latrines meer gericht zijn op FS-inzameling, wordt het legen, transporteren, behandelen en verwijderen / hergebruiken van de eindproducten vaak verwaarloosd. Als zodanig worden ze gekenmerkt door een "vul- en verlatingsstrategie" waarbij levensduur wordt gereguleerd door de beschikbaarheid van land - vooral met toenemende bevolking. Terwijl de verlaten latrines het gebied onbruikbaar maken voor andere winstgevende activiteiten, vindt het aldus begraven slib zijn weg naar ondiepe putten en grondwater als verontreiniging, en is het rijke organische materiaal niet meer beschikbaar voor biogeochemische cycli. Meer duurzame en kosteneffectieve interventies voor het beheer van fecaal slib, toepasbaar in informele stedelijke en voorstedelijke gebieden, zijn essentieel. De interventies moeten primair gericht zijn op de hele sanitatieketen: inzameling, lediging, transport, behandeling en definitieve verwijdering / eindgebruik.

Een op hulpbronnen gebaseerde benadering van sanitaire voorzieningen wordt steeds meer geïmplementeerd, vooral in wijken met hoge bevolkingsdichtheid en laag inkomen (HBLI- of sloppenwijken). Naast het verbeteren van de sanitaire voorzieningen zorgt deze benadering ervoor dat de nuttige stoffen die beschikbaar zijn in FS worden gebruikt voor verbetering van het menselijk leven, zoals biochemische energieterugwinning, organische mest, briketten en eiwitproductie. Deze aanpak is in Kenia overgenomen en praktisch toegepast door Sanergy, een sociale onderneming die werkt aan sanitaire voorzieningen in informele HBLI wijken. Ze passen het gebruik van urinescheidingstoiletten (*Urine Diverting Dehydrating Toilets*, UDDTs) toe in informele nederzettingen in Nairobi en registreren een dagelijkse verzameling van 8 ton ontlasting van UDDTs (UDDT-FS). Het UDDT-principe behelst een gescheiden inzameling van ontlasting en urine. Na toiletgebruik wordt er ongeveer 15 gram zaagsel toegevoegd om geur en vliegen te voorkomen. Na collectie wordt de UDDT-FS vervoerd naar een centrale verwerkingsinstallatie gelegen in Kinanie, 40 kilometer van Nairobi. Het afval ondergaat aanvullende behandelingsstappen, aangezien louter de toevoeging van zaagsel de aanwezige pathogenen niet voldoende inactiveert. Sanergy, dat compostering als belangrijkste verwerkingsmethode toepast, kan de grote toestroom van afval niet verwerken, en probeert daarom de behandelingsopties te diversifiëren.

Het sanitatiemodel van Sanergy is vrij succesvol. Eerdere pogingen om de sanitaire voorzieningen in HBLI wijken te verbeteren zijn vaak mislukt. Onze studie heeft tot doel deze mislukte pogingen te evalueren, en alternatieve technologieën te bieden voor de behandeling van FS, waarbij afval als grondstof wordt gezien en inactivering van pathogenen plaatsvindt. Het onderzoek is uitgevoerd in een aantal fasen:

a. Fase 1 – Beperkingen van de HBLI (hoge bevolkingsdichtheid, laag inkomen) wijkketen; case study informele wijk Kibera

b. Fase 2i & 2ii – Biochemische energieterugwinning en inactivatie van pathogenen tijdens anaërobe stabilisatie van UDDT-FS.

a) Fase 1 – Veldwerk

Gedurende het veldwerk zijn de beperkingen in de sanitaire voorzieningen in de informele wijken onderzocht. Resultaten worden gepresenteerd uit vragenlijsten en kwalitatieve interviews met belanghebbenden op alle niveaus van de sanitatieketen, d.w.z. inzameling, lediging, transport, verwijdering, behandeling en definitieve verwijdering / hergebruik. Het aanbieden van sanitaire voorzieningen is niet voldoende om de sanitatie-uitdaging in de informele wijken aan te pakken. De 'vul-en-verlaat' strategie kenmerkt het beheer van pitlatrines binnen de nederzetting.

Momenteel is er geen beleidskader voor planning, implementatie en beheer van sanitaire voorzieningen ter plaatse op stadsniveau. Dit leidt tot ongecoördineerde acties van de verschillende belanghebbenden in de sanitatieketen. Sanitaire aanbieders richten zich op het leveren van sanitaire voorzieningen maar dragen geen zorg voor het ledigen, transport, behandeling en verwijdering / hergebruik van eindproduct. Werkenden in de informele sanitatiesector hebben eveneens ongereglementeerde operaties: tot 85% van het opgehaalde fecaal slib (FS) komt onbehandeld in het milieu terecht. De betaal-per-gebruik aanpak van sanitaire voorzieningen verbetert de werking en het onderhoud van initiatieven: terwijl 73% van de gratis te gebruiken faciliteiten niet meer werd gebruikt nadat ze vol raakten, werden respectievelijk 89% en 77% van de gemeenschapsorganisaties (CBO's) en de door ondernemers beheerde faciliteiten goed beheerd. Verbeteringen op het gebied van op partnerschap gebaseerde sanitaire voorzieningen vormen een toegangspunt voor bredere initiatieven om de levensomstandigheden in informele nederzettingen te verbeteren. Door het betrekken van overheidsactoren, CBO's, gemeenschappen en werkenden in de informele sanitatiesector, heeft Umande Trust een economisch haalbare benadering van inclusieve duurzame sanitaire voorzieningen voor kansarme bevolking gecreëerd. De resultaten van het onderzoek zijn nuttig voor de lokale overheid en andere partners die betrokken zijn bij de verbetering van de sanitaire voorzieningen in de informele wijken.

b) Fase 2i & 2ii – Laboratorium- en pilotschaal experimenten

Alternatieve fecaal-slib (FS) behandelingstechnologieën zijn nodig voor het beheer van FS vanuit onsite-technologieën. Fase 2i en 2ii van onze studie verkenden het potentieel voor biochemische energierecuperatie en inactivatie van pathogenen tijdens anaërobe stabilisatie van UDDT-FS op i) laboratoriumschaal en ii) experimenten op pilootschaal.

2i. Laboratoriumtests

In deze studie is de mate van inactivering onderzocht van *Escherichia coli (E. coli)* en *Ascaris lumbricoides (A. lumbricoides)* eieren in fecaliën verkregen uit urinescheidingstoiletten (UDDT-FS) door hoge concentraties vluchtige vetzuren (VFA's) toe te voegen tijdens anaërobe omzetting. De impact van individuele VFA's op de inactivering van eieren van *E. coli* en *A. lumbricoides* in UDDT-FS werd beoordeeld door het toepassen van verschillende concentraties VFA in de vorm van acetaat, propionaat en butyraat. Hoge VFA concentraties werden ook bereikt door co-vergisting van UDDT-FS met organisch marktafval (OMW) in verschillende verhoudingen. Onderzoeksbevindingen toonden een positieve correlatie tussen *E. coli* logaritmische inactivatie en VFA-concentratie. Voor co-vergisting werd waargenomen dat de OMW-fractie in het voedingssubstraat de hoeveelheid VFA beïnvloedt, waarbij de hoeveelheid VFA toeneemt bij toenemende OMW-fractie. Een te hoge OMW-fractie is echter nadelig vanwege: 1) logistieke uitdagingen: inzameling, sortering en transportkosten van het afval van de sloppenwijken naar de verwerkingslocatie, en 2) bij een hoge OMW-fractie (UDDT-F:OMW ratio's van 1:4, 1:2 en 1:1) daalt de pH tot zeer lage niveaus, wat nadelig is voor de microbiële populatie, essentieel voor anaërobe vergisting. Als zodanig lijkt de ND-VFA-opbouw in het bereik van 1,2 - 1,8 meq / g totaal toegevoegde TS (*total solids*) voldoende te zijn voor voldoende inactivatie, wat overeenkomt met een ND-VFA-concentratie van ongeveer 2800-4300 mg/L. In 4 dagen werd tussen 3 en 5 *E. coli* log-inactivering bereikt bij een UDDT-F: OMW-verhouding van 2:1 tot 4:1. Als zodanig werd een UDDT-F: OMW-verhouding van 4:1 aanbevolen voor verdere evaluatie onder pilot scale onderzoek. Verder werd de biogasproductie bij de aanbevolen UDDT-F:OMW verhouding van 4: 1 geëvalueerd, waarbij ook twee controle-experimenten bij UDDT-F:OMW verhoudingen 1:0 en 0:1 werden ingesteld. De bevindingen toonden een hogere biogasproductie bij een hogere OMW-fractie, toegeschreven aan de hogere hydrolyse, opgemerkt in het vorige experiment. Echter, onder praktische toepassing kan vergisting van alleen OMW leiden tot verzuring van de reactor.

2ii. Experimenten op proefschaal

In het kader van experimenten op pilot schaal werden de vergisting en co-vergisting onderzocht van uitwerpselen uit urinescheidende toiletten (UDDT-FS) en gemengd organisch marktafval (OMW) in een eentraps en een tweetraps mesofiele pilot scale plugflow reactor. De gekozen UDDT-FS:OMW verhouding van 4:1 volgde uit de aanbevolen onderzoeksresultaten op laboratoriumschaal. Een controle-experiment op UDDT-FS: OMW-verhouding 1:0 werd ingesteld. Inactivatie van *E. coli*, de opbouw van vluchtige vetzuren (VFA) en de productie van biogas werden gemeten op bemonsteringspunten langs het reactorprofiel. De totale vaste stof (TS)-concentratie was gebaseerd op het maximum dat kon stromen zonder dat pompen nodig was, dus 12% TS werd toegepast. Een gehalte van 10% TS werd toegepast om het effect van inactivering van pathogenen bij een lagere TS-concentratie te beoordelen. Co-vergisting van UDDT-FS met OMW in een tweetraps reactor toonde 8,0 log pathogeeninactivering, hoger dan in de eentraps reactor met 5,7 loginactivering. De hoogste VFA-concentratie was 6,3 ± 1,3 g/L, verkregen bij een pH van 4,9 in de hydrolyse- / acidogenesereactor onder toepassing van een UDDT-FS: OMW-verhouding van 4: 1 en 12% TS, wat overeenkomt met een niet-gedissocieerde (ND)-VFA-concentratie van 6,9 ± 2,0 g / L en vervalsnelheid van 1,6/d. In de daaropvolgende methanogene plug-flow reactor werd een vervalsnelheid van 1,1/d bereikt binnen het eerste derde deel van de reactorlengte, afnemend tot 0,6/d binnen het laatste derde deel van de reactorlengte.

In daaropvolgende proefschaalexperimenten werden eentraps en tweetraps plugstroomvergisters onderzocht, waarbij de UDDT-FS:OMW-verhouding van 4:1 en 1:0 werd toegepast bij 10% en 12% TS-concentraties. Vergelijkbare methaanproductie werd waargenomen in eentraps (R_s-4:1, 12%) (314 ± 15 ml CH_4/g VS toegevoegd) en tweetraps vergisters (R_{am}-4:1, 12%) (325 ± 12 ml CH_4/g VS toegevoegd), bij toepassing van een slurryconcentratie van 12% totaal vaste stoffen (TS). De biogasproductie in R_{am}-4: 1, 12% vergister (571 ± 25 ml CH_4/g VS toegevoegd) was echter ongeveer 12% hoger dan in de R_s-4:1, 12%, dus aanzienlijk meer. Dit werd toegeschreven aan verbeterde oplosbaarheid van afval en verhoogde CO_2-oplossing, als gevolg van het mengen van het bicarbonaatrijke methanogene effluent met het influent met lage pH (4,9) afkomstig van de fase van voorverzuring, met als doel neutralisatie. Een hogere processtabiliteit werd waargenomen in de eerste delen van de tweetraps vergister met propstroom, gekenmerkt door lagere VFA-concentraties.

De onderzoeksresultaten tonen het potentieel aan voor een tweetraps plug-flow reactor als een kosteneffectieve FS-behandelingstechnologie. Voor praktische doeleinden beveelt onze studie aan om de reactor te laten werken met een UDDT-FS: OMW-verhouding van 4:1, bij een TS-concentratie van 12%, 35^{0}C, waardoor de UDDT-FS-fractie die per tijdseenheid kan worden behandeld, maximaal is. De TS-concentratie van 12% was de hoogst mogelijke concentratie die kon stromen zonder mechanisch pompen. Verhoogde inactivering van pathogenen werd waargenomen bij een toenemende OMW-fractie in voedingssubstraat, maar kan leiden tot procesfalen als gevolg van zeer lage pH-waarden, die toxisch kunnen zijn voor *E. coli*, evenals voor andere microbiële populaties. Het systeem bevordert niet alleen de hygiënische behandeling van fecaliën, maar maakt ook het terugwinnen van hulpbronnen mogelijk. Biogas dat uit de faciliteiten wordt geproduceerd, wordt verkocht aan de HBLI-bewoners, als brandstof voor koken in de centra of om warme douches te leveren, tegen betaling afhankelijk van het soort aangeboden service en varieert van 0,2 euro (rijst koken of warme douche) tot 0,4 euro (het koken van bonen). Daarnaast kan het effluent worden gebruikt als alternatieve mestbron om de landbouwproductie te verhogen, ter vervanging van commerciële mestbronnen.

Bovendien heeft het systeem potentieel voor toepassing als off-site FS-behandelingstechnologie op elke schaal, dus toepasbaar voor FS-behandeling in HBLI in Sub-

Sahara Afrika. Bij een off-site systeem wordt afval elders gegenereerd en afgeleverd voor verwerking op een centrale verwerkingslocatie. Op basis hiervan kan de technologie worden toegepast om fecaal afval te verwerken dat is ingezameld met andere technologieën, b.v. UDDT, peepoo-zakjes etc. Om de logistieke en operationele kosten van het transport van UDDT-FS te verlagen, moeten off-site verwerkingslocaties zo dicht mogelijk bij de UDDT-FS-inzamelpunten worden geplaatst.

Table of Contents

Chapter 1: Introduction

1.1   General introduction

Worldwide, 2.7 billion people are using onsite sanitation systems, with number expected to increase to 5 billion by 2030 (Strande *et al.*, 2014). This trend is expected due to rapid population growth and emergence of LIHDS in urban cities of developing countries. In Kenya, rapid urbanisation has brought influx of people to urban centres. Owing to a lag in planning and development of infrastructure to meet the demands of the growing population, more than 100 LIHDS emerged in Nairobi, Kenya (AWF, 2013). The settlements are characterised by haphazard development, poor sanitation, high population, high poverty levels and insecure land tenure (Katukiza *et al.*, 2010; Kulabako *et al.*, 2010; Mels *et al.*, 2009; Scott *et al.*, 2013).

Previous attempts to improve sanitation applied sewer-based sanitation system which are expensive to develop in poor and less developed countries (Lalander *et al.*, 2013; Mara, 2013), in addition to significant costs for maintaining and upgrading infrastructure (Kone, 2010; Schertenleib, 2005; Zimmer & Hofwegen, 2006). Moreover, haphazard development and challenges of informal settlements (Katukiza *et al.*, 2010; Kulabako *et al.*, 2010; Mels *et al.*, 2009; Scott *et al.*, 2013) would make pipe layout for sewer-based systems a daunting task. In addition, 75% of the residents buy water from kiosks at prices far higher than those paid by middle and high-income households, further limiting connection to the city's sewer system (UN-Habitat, 2006).

Onsite sanitation systems, previously viewed as sanitation technologies for rural areas have largely been adopted; globally, 2.7 billion people currently rely on on-site sanitation (Strande *et al.*, 2014). In Nairobi, only 12% of the population is connected to conventional sewer system (MOH, 2016), with 78% relying on onsite sanitation systems. Despite several organisations spearheading provision of onsite sanitation systems in Nairobi, sanitation status remain low (Losai., 2011). The systems, if not well managed can lead to high environmental and public health risks (Winblad & Simpson-Hebert, 2004), owing to high pathogenic load in excreta (Feachem *et al.*, 1983).

In Kenya, consequences of poor sanitation are glaring. Unsafe water and sanitation (5.3%) is the second leading risk factor and contributor to all mortality (deaths) and morbidity burden in Kenya (WHO, 2009). KES 27 billion (USD 365 million) is spent annually by Kenyan government (one percent of national gross domestic product (GPD)) in the treatment of sanitation related illnesses (MOH, 2016). Among children, diarrhoeal disease and intestinal worm infestation contribute to at least 40% of deaths among under five children (MOH, 2010). In addition to 35% of children suffering from moderate to severe stunting (UNICEF, 2013), with LIHDS being worst hit off by the sanitation menace (Gulis *et al.*, 2004; Mberu *et al.*, 2016; Zulu *et al.*, 2011).

Even with adoption of low cost, water saving sanitation facilities such as UDDT-FS for sanitation improvement in informal settlements, the health risk associated to excreta management are not eliminated. The addition of saw dust or ash after toilet use is not sufficient to kill pathogens (Niwagaba *et al.*, 2009a). Thus, an extra pathogen inactivation step is required after waste collection, especially when the faecal matter will be valorised for agricultural purposes and also avoid contamination of water bodies. This calls for research more of technological options for management of FS generated from onsite sanitation systems (Strande *et al.*, 2014).

Anaerobic digestion (AD) offers an attractive approach in FS treatment (Rajagopal *et al.*, 2013) as its effluent has good fertilising qualities in addition to biochemical energy recovery through methane build-up (Avery *et al.*, 2014; Fonoll *et al.*, 2015; Nallathambi Gunaseelan, 1997; Romero-Güiza *et al.*, 2014). Moreover, recovery of methane from organic waste reduces global warming and offers an alternative to fossil fuels (Abbasi *et al.*, 2012). However, when applying UDDT-FS as sole substrate, application of AD for treatment of FS has been limited by unsatisfactory pathogen inactivation (Chaggu, 2004; Dudley *et al.*, 1980; Foliguet & Doncoeur, 1972; Leclerc & Brouzes, 1973; McKinney *et al.*, 1958; Pramer *et al.*, 1950) in addition to low methane production (Rajagopal *et al.* 2013; Fagbohungbe *et al.* 2015). Microbiological safety of the digestate and treated sludge is essential especially during reuse for agricultural purposes/disposal to environment (Avery *et al.* 2014), as it can lead to transmission of enteric diseases (Pennington, 2001; Smith *et al.*, 2005). As such, enhanced pathogen inactivation during AD was explored in this study.

This study conducted in two phases explored LIHDS sanitation chain, linking sanitation provision to working FSM value chain. Phase 1 explored impediments to LIHDS sanitation enhancement during system provision, emptying, transportation, treatment, disposal/ reuse. In this, questionnaires were administered to key stakeholders in LIHDS sanitation provision and semi-structured interviews organised with government representatives. In Phase 2 of the study the proposed system was aimed at enhancing pathogen inactivation and biochemical energy recovery during anaerobic digestion of UDDT-FS. In this, natural build-up of non-dissociated volatile fatty acids (ND-VFA) acts as the sanitising agent. ND-VFAs pass freely through bacterial cell walls by passive diffusion and affect the internal pH causing inactivation (Jiang *et al.*, 2013; Wang *et al.*, 2014; Zhang *et al.*, 2005; Riungu *et al.*, 2018). In our two-stage reactor setup, ND-VFA build-up is enhanced by co-digestion of human waste and organic market waste (OMW). OMW by virtue of containing readily degradable organic fraction undergoes rapid hydrolysis leading when used as co-substrate medium resulting to increase in concentrations of ND-VFA in the digestion medium (Riungu *et al.*, 2018a; Zhang *et al.*, 2008; Zhang *et al.*, 2005). Enhanced build-up of ND-VFA concentrations during co-digestion of sewage sludge and other organic waste can be achieved by inhibition of methanogenesis (Wang *et al.*, 2014), through use of a two-stage reactor system, where hydrolysis/or acidogenesis and methanogenesis are separated. The different species of micro-organisms involved in the AD process can be divided into two main groups of bacteria, namely organic acid producing and organic acid consuming or methane forming microorganisms (Rincón *et al.*, 2008). They operate under different pH conditions: whereas the optimal pH for acidogenic bacteria activity ranges between 5 and 7 (Fang & Liu, 2002; Guo *et al.*, 2010; Liu *et al.*, 2006; Noike *et al.*, 2005), methanogenic activity requires a minimum pH of 6.5 (Wang *et al.*, 2014b; Yuan *et al.*, 2006). The increased solubilisation of waste in the hydrolysis reactor leads to higher biogas production in the methanogenic stage. The key drawback, high VFA concentration in the acidogenic reactor which requires pH correction for stable methanogenesis (Zuo *et al.,* 2014), is counteracted by use of a recycle stream for pH regulation.

1.2 Scope of research

This PhD thesis explored LIHDS sanitation chain, linking sanitation provision to working FSM value chain. sought to enhance pathogen inactivation and biogas production during anaerobic stabilisation of UDDT-FS. The research was conducted in two main phases;

Phase 1; Field work-evaluation of the limitations of LIHDS sanitation chain; Case study-Kibera LIHDS

Phase 2; Consisted of two parts 2i)-laboratory scale experiments and 2ii) -pilot scale experiments. Phase 2i laboratory experimental results were applied for setting up pilot scale experiments in phase 2ii.

a. Laboratory scale tests.

The potential of *Escherichia coli (E. coli)* and *Ascaris lumbricoides (A. lumbricoides)* eggs inactivation in faecal matter coming from urine diverting dehydrating toilets (UDDT-FS) by applying high concentrations of volatile fatty acids (VFAs) during anaerobic digestion was examined. The impact of individual VFAs on *E. coli* and *A. lumbricoides* eggs inactivation in UDDT-FS was assessed by applying various concentrations of store-bought acetate, propionate and butyrate. VFA build-up was also obtained by performing co-digestion of UDDT-FS with organic market waste (OMW) using various mixing ratios. Biochemical Methane Potential tests (BMP) experiments were also conducted under laboratory conditions to evaluate the potential for biochemical energy recovery from urine diverting dehydrating toilets faeces (UDDT-FS). From phase 2i findings, a practical UDDT-FS:OMW mix ratio was established and recommended for further test under pilot scale.

b. Pilot scale tests

Digestion and co-digestion of faecal matter collected from Urine Diverting Dehydrating Toilet Faeces (UDDT-FS) and mixed Organic Market Waste (OMW) was studied in single stage pilot scale mesophilic plug-flow anaerobic reactors at UDDT-FS:OMW ratios 4:1 and 1:0. *Escherichia coli (E. coli)* inactivation and Volatile Fatty Acids (VFA) build-up was monitored at sampling points located along the reactor profile. A two-stage reactor was assessed, applying a UDDT-FS:OMW ratio of 4:1 and 10% or 12% TS slurry concentrations. In subsequent trials, performance of single- stage and two stage reactors comprising an acidogenic reactor and a methanogenic reactor, in terms of biogas production was evaluated. Test substrates used were UDDT-FS: organic market waste (OMW) ratios 4:1 and 1:0, at 10% and 12% total solids (TS) substrate concentration.

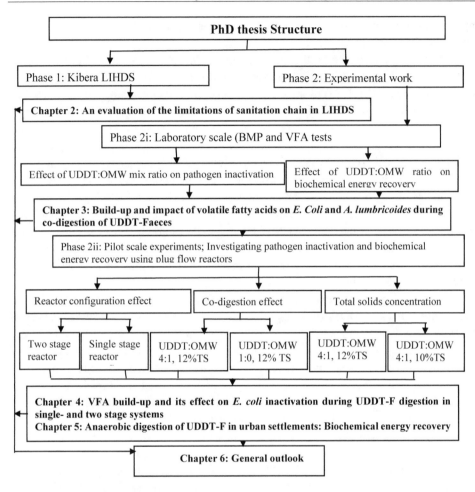

Figure 1.1: PhD thesis structure

1.3   Thesis outline

This thesis is organised into 6 chapters (Figure 1.1): introduction (Chapter 1), evaluations of the limitations of LIHDS sanitation chain ((Chapter 2), 3 research chapters (chapter 3, 4 and 5) as well as general outlook (Chapter 6). The introductory part gives a justification of study topic, scope and outline of the research. The second chapter evaluates the limitations of LIHDS sanitation, case study Kibera. Whereas the background information captures a brief background on the sanitation status in Kenya's low income, high density settlements (LIHDS), the institutional arrangements for urban sanitation in Kenya and impediments to provision of LIHDS sanitation elucidates why sanitation remains poor even with efforts for improvement.

Chapter three examines the potential of *Escherichia coli (E. coli)* and *Ascaris lumbricoides (A. lumbricoides)* eggs inactivation in faecal matter coming from urine diverting dehydrating toilets (UDDT-FS) by applying high concentrations of volatile fatty acids (VFAs) during anaerobic digestion. The impact of individual VFAs on *E. coli* and *A. lumbricoides* eggs inactivation in UDDT-FS was assessed by applying various concentrations of store-bought acetate, propionate and butyrate. High VFA concentrations were also obtained by performing co-digestion of UDDT-FS with organic market waste (OMW) using various mixing ratios. Study findings showed an increase in *E. coli* log inactivation at increasing store bought VFA concentrations. In co-digesting UDDT-FS and organic market waste (OMW), *E. coli* and *A. lumbricoides* egg inactivation was found to be related to the concentration of non-dissociated VFA (ND-VFA), increasing with the OMW fraction in the feed substrate. The findings led to an *E. coli* log inactivation to below detectable levels and complete *A. lumbricoides* egg inactivation in less than four days at the recommended UDDT-FS:OMW ratio 4:1 and ND-VFA concentration of 4800-6000 mg/L.

Chapter 4 evaluates pathogen inactivation during digestion and co-digestion of faecal matter collected from UDDTs (UDDT) and OMW. Experiments were conducted at UDDT-FS:OMW ratio 4:1, at 10% and 12% total solids (TS) concentrations. The UDDT-FS:OMW mix ratio selection was based on a series of laboratory scale batch-tests derived experimental data on the effect of substrate concentration on pathogen inactivation (Chapter 1) whereas 12% TS was based on treating the highest possible TS concentration that could freely flow through the plug-flow reactor without the necessity of using pumps; 10% TS was based on the need to investigate pathogen removal at a lower TS concentration. Study findings showed that co-digesting UDDT-FS with OMW in two stage reactor achieved higher pathogen inactivation (8.0 log inactivation) than corresponding single stage reactor (5.7 log inactivation). In addition, single stage anaerobic digestion of UDDT-FS achieved higher pathogen inactivation with a mixed substrate (5.7 log inactivation) than during single stage substrate digestion of UDDT-FS (3.4 log inactivation).

Chapter 5 assessed biochemical energy recovery during digestion and co-digestion of faecal matter collected from urine diverting dehydrating toilet faeces (UDDT-FS) and mixed organic market waste (OMW). Laboratory scale biochemical methane potential (BMP) tests showed an increase in methane production with an increased OMW fraction in the feed substrate.

In subsequent pilot scale experiments, single and two stage plug flow digester were assessed, applying UDDT-FS: OMW ratio of 4:1 and 1:0, at 10 or 12% TS slurry concentrations. Comparable methane production was observed in single stage ($R_{s-4:1, \, 12\%}$) (314 ± 15 mL CH_4 /g VS added) and two stage ($R_{am-4:1, \, 12\%}$) (325 ± 12 mL CH_4 /g VS added) digesters, when applying 12% total solids (TS) slurry concentration. However, biogas production in $R_{am-4:1, \, 12\%}$ digester (571 ± 25 mL CH_4/g VS added) was about 12% higher than in the $R_{s-4:1, \, 12\%}$, significantly more than the slight difference in methane production, i.e. 3-4%.

6

This was attributed to enhanced waste solubilisation and increased CO_2 dissolution, resulting from mixing the bicarbonate-rich methanogenic effluent for neutralisation purposes with the low pH (4.9) influent coming from the pre-acidification stage. Higher process stability was moreover observed in the first parts of the plug flow two- stage digester, characterised by lower VFA concentrations.

Chapter 6 provides a summary of key achievements and recommendations arising from the study. In addition, the Chapter assess sustainability of anaerobic digestion as applied for sanitation improvement in Kenya.

References

Abbasi, T., Tauseef, S.M., Abbasi, S.A. 2012. Anaerobic digestion for global warming control and energy generation - an overview. Renew. Sust. Energy Rev. 16, 3228-3242.

Avery, L.M., Anchang, K.Y., Tumwesige, V., Strachan, N., Goude, P.J. 2014. Potential for Pathogen reduction in anaerobic digestion and biogas generation in Sub-Saharan Africa. *Biomass and Bioenergy*, 70(0), 112-124.

AWF. 2013. Expanding branded toilet entrepreneurship for improved sustainable sanitation in poor neighbourhoods of Nairobi, Kenya.

Chaggu, E.J. 2004. Sustainable Environmental Protection Using Modified Pit-Latrines. Ph.D Thesis, Wageningen University, The Netherlands.

Dudley, D.J., Guentzel, M.N., Ibarra, M.J., Moore, B.E., Sagik, B.P. 1980. Enumeration of potentially pathogenic bacteria from sewage sludge. *Applied Environmental Microbiology*, 39, 118-126.

Fang, H.H.P., Liu, H. 2002. Effect of pH on hydrogen production from glucose by a mixed culture. *Bioresource Technology*, 82(1), 87-93.

Feachem, R.G., Bradley, D.J., Garelick, H., D., M.D. 1983. Sanitation and Disease Health Aspects of Excreta and Wastewater Management. *Report No.:11616 Type: (PUB)*

Foliguet, J.M., Doncoeur, F. 1972. Inactivation in fresh and digested wastewater sludges by pasteurization. *Water Research*, 6, 1399-1407.

Fonoll, X., Astals, S., Dosta, J., Mata-Alvarez, J. 2015. Anaerobic co-digestion of sewage sludge and fruit wastes: Evaluation of the transitory states when the co-substrate is changed. *Chemical Engineering Journal*, 262(0), 1268-1274.

Gulis, G., Mulumba, J.A.A., Juma, O., Kakosova, B. 2004. Health status of people of slums in Nairobi, Kenya. *Environmental Research*, 96(2), 219-227.

Guo, X.M., Trably, E., Latrille, E., Carrère, H., Steyer, J.-P. 2010. Hydrogen production from agricultural waste by dark fermentation: A review. *International Journal of Hydrogen Energy*, 35(19), 10660-10673.

Katukiza, A.Y., Ronteltap, M., Oleja, A., Niwagaba, C.B., Kansiime, F., Lens, P.N.L. 2010. Selection of sustainable sanitation technologies for urban slums — A case of Bwaise III in Kampala, Uganda. *Science of The Total Environment*, 409(1), 52-62.

Kone, D. 2010. Making urban excreta and wastewater management contribute to cities economic development: a pradigm shift *Water Policy*, 12, 602-610.

Kulabako, N., Nalubega, M., Wozei, E., Thunvik, R. 2010. Environmental health practices, constraints and possible interventions in peri-urban settlements in developing countries-a review of Kampala, Uganda. *Int J Environ Health Res* 20(4), 231-257.

Lalander, C.H., Hill, G.B., Vinnerås, B. 2013. Hygienic quality of faeces treated in urine diverting vermicomposting toilets. *Waste Management*, 33(11), 2204-2210.

Leclerc, H., Brouzes, P. 1973. Sanitary aspects of sludge treatment. *Water Research*, 7(3), 355-360.

Liu, D., Liu, D., Zeng, R.J., Angelidaki, I. 2006. Hydrogen and methane production from household solid waste in the two-stage fermentation process. *Water Research*, 40(11), 2230-2236.

Losai. 2011. Landscape Analysis and Business Model Assessment in Fecal Sludge Management: Extraction and Transportation Models in Africa.

Mara, D. 2013. Pits, pipes, ponds – And me. *Water Research*, 47(7), 2105-2117.

Mberu, B.U., Haregu, T.N., Kyobutungi, C., Ezeh, A.C. 2016. Health and health-related indicators in slum, rural, and urban communities: a comparative analysis. African Population and Health Research Center, Nairobi, Kenya.

McKinney, R.E., Langley, H.E., Tomlinson, H.D. 1958. Survival of Salmonella typhosa during anaerobic digestion. I. Experimental methods and high rate digester studies. *Sewage Ind. Wastes*, 30, 1467-1477.

Mels, A., Castellano, D., Braadbaart, O., Veenstra, S., Dijkstra, I., Meulman, B., Singels, A., Wilsenach, J.A. 2009. Sanitation services for the informal settlements of Cape Town, South Africa. *Desalination*, 248(1–3), 330-337.

MOH. 2016. Kenya Environmental Sanitation and Hygiene Policy 2016-2030.

MOH. 2010. Review of the Kenya Health Policy Framework, 1994 – 2010.

Nallathambi Gunaseelan, V. 1997. Anaerobic digestion of biomass for methane production: A review. *Biomass and Bioenergy*, 13(1–2), 83-114.

Niwagaba, C., Kulabako, R.N., Mugala, P., Jönsson, H. 2009. Comparing microbial die-off in separately collected faeces with ash and sawdust additives. *Waste Management*, 29(7), 2214-2219.

Noike, T., Ko, I., Yokoyama, S., Kohno, Y., Li, Y. 2005. Continuous hydrogen production from organic waste. *Water Sci Technol*, 52(1-2), 145-51.

Pennington, T.H. 2001. Pathogens in agriculture and the environment. In: Pathogens in Agriculture and the Environment, Meeting organised by the SCI Agriculture and Environment Group, 16 October, SCI, London.

Pramer, D., H. , Heukelekian, Ragotskie, R.A. 1950. Survival of tubercule bacilli in various sewage treatment processes. I. Development of a method for the quantitative recovery of mycobacteria from sewage. *Public Health Rep.* , 65, 851-859.

Rajagopal, R., Lim, J.W., Mao, Y., Chen, C.-L., Wang, J.-Y. 2013. Anaerobic co-digestion of source segregated brown water (feces-without-urine) and food waste: For Singapore context. *Science of The Total Environment*, 443(0), 877-886.

Rincón, B., Sánchez, E., Raposo, F., Borja, R., Travieso, L., Martín, M.A., Martín, A. 2008. Effect of the organic loading rate on the performance of anaerobic acidogenic fermentation of two-phase olive mill solid residue. *Waste Management*, 28(5), 870-877.

Riungu, J., Ronteltap, M., van Lier, J.B. 2018. Build-up and impact of volatile fatty acids on *E. coli* and *A. lumbricoides* during co-digestion of urine diverting dehydrating toilet (UDDT-F) faeces. *J Environ Manage*, 215, 22-31.

Romero-Güiza, M.S., Astals, S., Chimenos, J.M., Martínez, M., Mata-Alvarez, J. 2014. Improving anaerobic digestion of pig manure by adding in the same reactor a stabilizing agent formulated with low-grade magnesium oxide. *Biomass and Bioenergy*, 67, 243-251.

Schertenleib, R. 2005. From conventional to advanced environmental sanitation. *Water Science & Technology*, 51(10), 7-14.

Scott, P., Cotton, A., Sohail Khan, M. 2013. Tenure security and household investment decisions for urban sanitation: The case of Dakar, Senegal. *Habitat International*, 40(0), 58-64.

Smith, S.R., Lang, N.L., Cheung, K.H.M., Spanoudaki, K. 2005. Factors controlling pathogen destruction during anaerobic digestion of biowastes. *Waste Management*, 25(4), 417-425.

Strande, L., Ronteltap, M., Brdjanovic, D. 2014. Faecal Sludge Management Systems Approach for Implementation and Operation. IWA publishing, ww.iwapublishing.com.

UN-Habitat. 2006. Nairobi urban sector profile; Rapid Urban Sector Profiling for Sustainability (RUSPS). UNICEF. 2013. The State of the World's Children(SOWC): Children with Disabilities, UNICEF, NewYork, May 2013.

Wang, K., Yin, J., Shen, D., Li, N. 2014. Anaerobic digestion of food waste for volatile fatty acids (VFAs) production with different types of inoculum: Effect of pH. *Bioresource Technology*, 161, 395-401.

WHO, 2009. Global Health Risks Global Health Risks, WHO Mortality and burden of disease attributable to selected major risks.

Winblad, U., Simpson-Hebert, M. 2004. *Ecological sanitation - revised and enlarged edition*. Stockholm Institute of Environment, Stockholm, Sweden.

Yuan, H., Chen, Y., Zhang, H., Jiang, S., Zhou, Q., Gu, G. 2006. Improved Bioproduction of Short-Chain Fatty Acids (SCFAs) from Excess Sludge under Alkaline Conditions. *Environmental Science & Technology*, 40(6), 2025-2029.

Zhang, B., He, P., LÜ, F., Shao, L. 2008. Enhancement of anaerobic biodegradability of flower stem wastes with vegetable wastes by co-hydrolysis. *Journal of Environmental Sciences*, 20(3), 297-303.

Zhang, B., Zhang, L.L., Zhang, S.C., Shi, H.Z., Cai, W.M. 2005. The influence of pH on hydrolysis and acidogenesis of kitchen wastes in two-phase anaerobic digestion. *Environmental technology*, 3, 329-339.

Zimmer, D., Hofwegen, P. 2006. Costing MDG Target 10 on water supply and sanitation: comparative analysis, obstacles and recommendations.

Zulu, E.M., Beguy, D., Ezeh, A., C.,, Bocquier, P., Madise, N.J., Cleland, J., Falkingham, J. 2011. Overview of migration, poverty and health dynamics in Nairobi City's slum settlements. *Journal of Urban Health: Bulletin of the New York Academy of Medicine, Vol. 88, Suppl. 2 doi:10.1007/s11524-011-9595-0.*

Zuo, Z., Wu, S., Zhang, W., Dong, R. 2014. Performance of two-stage vegetable waste anaerobic digestion depending on varying recirculation rates. *Bioresource Technology*, 162, 266-272.

Chapter 2: An evaluation of the limitations of sanitation chain in low income, high density settlements: Case study, Kibera, Kenya.

This Chapter is based on a paper: Riungu J., Ronteltap, M., van Lier, J.B. (Under internal review). An evaluation of the limitations of sanitation chain in low income, high density settlements: Case study, Kibera, Kenya

Abstract

Achieving an operational onsite sanitation chain requires well-managed services in all aspects of excreta management: collection, emptying, transport, treatment and disposal/ reuse. To cover the full chain, however, remains a difficult challenge, and is even harder in informal housing areas. This paper presents an overview of the key limitations to sanitation improvement in low income, high density settlements (LIHDS). In particular, it focuses on the sanitation situation in Kibera LIHDS, Nairobi, Kenya. Results are presented from questionnaires and qualitative semi-structured interviews with stakeholders involved in sanitation provision within the settlement.

Currently, there is no policy framework governing planning, implementation and management of onsite sanitation at City level. This leads to uncoordinated actions among the various stakeholders along the sanitation chain. Sanitation providers focus on provision of sanitation facilities, neglecting emptying, transportation, treatment and disposal/reuse of end product. Pit emptiers likewise have unregulated operations with faecal sludge (FS) ending up untreated into the environment (85%).

Pay-and-use approaches to sanitation provision enhances the effectiveness of operation and maintenance initiatives: whereas in 73% free-to-use facilities were abandoned on fill-up, 89% and 77% of community-based organisation- (CBOs) and entrepreneur-managed pay-and-use facilities respectively were well managed. Partnership-based sanitation provision improvements provide an entry point for broader initiatives to improve living conditions in informal settlements. By involving government actors, CBOs, community, and pit emptiers, Umande Trust has created an economically viable approach of inclusive sustainable sanitation for underprivileged population. Results from the study are useful to the local government and other partners involved in sanitation improvement within the settlement.

Keywords: Sanitation, faecal sludge, disabling environment, informal settlement

2.1   Introduction

The proper implementation of sanitation measures to safeguard human health has imperative public health benefits, including improved human dignity, safety, health, and well-being. Among the 2.7 billion people globally served by onsite sanitation technologies (Strande *et al.*, 2014) and the nearly 892 millions of people practicing open defecation (Saleem *et al.*, 2019), treatment of faecal sludge generated remains a challenge. Onsite technologies encompasses all technologies where human excreta collection, storage and treatment (where this exists) are contained within the place occupied by the dwelling and its immediate surroundings (KESSF, 2016).

Engineers and policy makers previously viewed waterborne, sewer-based systems as the most viable long-term solution to fulfil sanitation needs, with onsite technologies being perceived as temporary solutions. However, to date, onsite sanitation technologies have been largely adopted as a sanitation solution in urban areas in developing countries (65-100%) (Strande *et al.*, 2014). When zooming in on Kenya, only 12% of the 40 million population is served by a sewer system of which only 5% is effective, whereas among the poor urban settlements, less than 20% have access to any form of sanitation (MOH, 2016). The high adoption of onsite technologies accompanied by a lag in technological development for management of the faecal sludge collected in these facilities has led to a sanitation crisis (Kenya Report, 2011), key being the increase in sanitation related diseases. Globally, diarrheal disease is the 9th largest cause of death among all ages and the 4th leading cause of mortality among children under the age of 5 years, whereas in sub-Saharan Africa, over 40% of all-age deaths and approximately 60% of under-5 mortality are linked to diarrheal diseases (Troeger *et al.*, 2017; Winter *et al.*, 2019).

Goal number 6 of the SDGs is to "Ensure availability and sustainable management of water and sanitation for all", where Target 6.2 aims by 2030, to achieve access to adequate and equitable sanitation and hygiene for all and end open defecation, paying special attention to the needs of women and girls and those in vulnerable situations (UN-SDG, 2015). The route to achieve this is by addressing the whole sanitation chain, so that one can ensure that faecal sludge is effectively and hygienically managed from collection to disposal/reuse. As such, an elaborate FS management system is essential for the resulting faecal sludge accumulating in onsite sanitation technologies (Schouten & Mathenge, 2010; AECOM and Sandec, 2010; Katukiza *et al.*, 2010). However, planning a complete FS management system especially in LIHDS in developing countries still remains a challenge.

Nairobi, Kenya's capital city, is home to more than 100 unplanned LIHDS (AWF, 2013). Despite occupying only 5% of the total residential area (NCWSC/AWSB, 2009), the settlements house 60% of Nairobi population (3.1 million (KNBS, 2009)). The settlements are not acknowledged by the state and local planning authorities (Huchzermeyer & Karam, 2006; K'Akumu & Olima, 2007; Syagga, 2011; UN-Habitat, 2006) thus crucial infrastructure and basic services such as water and sanitation are lacking (Wamuchiru, 2015). Moreover, they have unique institutional, demographic, socio-economic and environmental challenges that are context specific (Hogrewe et al., 1993; Lüthi et al., 2009; Okurut et al., 2015). As such, provision of sanitation in these areas is complex; the required sanitation services include: construction and installation of sanitation facilities, supply of sanitation products, repair and maintenance of facilities, emptying services, transportation, treatment, and safe disposal of waste, as well as education and/or sensitisation of the community on hygienic practices (Okurut et al., 2015). Efforts towards provision of sanitation have focused on construction and installation of facilities, such as pit latrines, ventilated improved pit latrines, and pour flush toilets connected to septic tanks and/or pits (Gulyani & Talukdar, 2008; Schouten & Mathenge, 2010; Wasao, 2002). Challenges faced in sanitation provision are that where facilities exist, they may either be abandoned, misused or never used at all (Mara et al., 2010) or shared (Tumwebaze et al., 2013), not clean and/or adequate enough to provide dignity and privacy (Okurut et al., 2015; Van Der Geest, 2002).

Consideration of local sanitary demands (customs and habits) related to sanitation and human waste handling is critical in ensuring that sanitation interventions facilitate the realisation of public health benefits (Evans & Tremolet, 2010), (Mosler, 2012; Peal et al., 2010). With poor understanding of local sanitary demands, communities may proceed with traditional excreta disposal practices e.g. founded on traditional beliefs and cultural influences and are unwilling to use systems that conflict with these (Van der Hoek et al., 2010). Proper consideration of both hardware (human interface) and local demands regarding sanitation services within LIHDS necessitates a well-balanced coordination of all stakeholders involved, having the intended users at the core of focus (Okurut et al., 2015). These stake include; government representatives, financers, service providers, consumer representatives, land management entities and health sector promoters acting at any point along the sanitation service chain. Proper coordination facilitates service provision that links the various sections of the sanitation chain, i.e. collection, transportation, treatment, final disposal/reuse, identifying any possible weak link that may lead to failed attempts of sanitation enhancement.

This study evaluates the impediments to sanitation improvement in LIHDS; case study Kibera, Kenya. It explores the prevailing sanitation chain, trying to elucidate the key areas requiring interventions and identifying the limiting aspects in LIHDS sanitation within Kibera settlement. Main research objectives were to: (i) Assess the sanitation chain within the settlement; (ii) Study and map the applied desludging techniques and faecal sludge management options; (iii) Evaluate the challenges faced by the stakeholders in sanitation provision. The results of the study may assist government and other organisations working on sanitation provision in LIHDS.

2.2 Methods

2.2.1 Introduction

Kibera LIHDS is located approximately 7 km South West of Nairobi, the capital city of Kenya. Kibera is administratively subdivided in 12 villages. The field work for this research was conducted by addressing various key stakeholders in the sanitation chain. In order to obtain the required data, several methods were used:
(i)   questionnaires were distributed and semi-structured interviews were carried out;
(ii)   observations were done by random walks in the settlement area;
(iii)   documents from NGOs and governmental bodies were studied
English or Kiswahili languages were used during administration of the questionnaires as well as during conducting interviews. In addition, face to face interviews were held to gather information from tenants in plots where the sanitation facilities were found to be full of excreta. All data collected was analysed using SPSS software and MS-excel. The field work was conducted for a period of five months commencing January 2014.

2.2.2   Questionnaires

Information was elicited by use of questionnaires from individuals involved in various sanitation chain aspects within the settlement. These included sanitation providers, pit emptiers and sanitation workers.

i.   Sanitation providers- This group involved the organisations/ individuals responsible for provision of sanitation facilities and included NGOs, landlords, entrepreneurs and churches. Questionnaires were administered to 63 sanitation providers. A total of 63 were interviewed.

ii.   Pit emptiers- This group involved people responsible for emptying of sludge from the sanitation facilities. The study targeted the maximum number of desludging groups possible, which highly depended on the willingness to provide information. Questionnaires were administered to 80 pit emptiers.

iii.   Sanitation workers- This group involved individuals responsible for daily maintenance and operation (M&O) of the sanitation facilities. The questionnaires were administered to sanitation workers manning 600 sanitation facilities that were visited during the study.

iv.   Semi-structured interviews- Semi-structured interviews were conducted with key actors in faecal sludge management (FSM) at city level. They included: Nairobi City Water and Sewerage Company (NCWSC), Nairobi City Council (NCC) and National Environment and management Authority (NEMA). The interviews revolved around their establishment and role in sanitation provision in LIHDS and challenges experienced. In addition, rights-based organisation (Umande Trust) and Sanergy, working on sanitation enhancement within the settlement was interviewed.

2.3   Results

2.3.1   Institutional and Legal Framework for FSM in Nairobi City

Insight in the institutional and legal framework was obtained from literature and discussions with key stakeholders that govern sanitation provision at City level. Literature and reports identified included: The Nairobi City Water and Sewerage Company (NCWSC, 2009), Environmental Management Coordination Act (EMCA) of 1999, Kenya Environmental Hygiene and Sanitation Strategic Framework (KESSF, 2016-2020), and Water Services Regulatory Board report (WASREB, 2016). In addition, interviews were held with governments' representatives in FSM at city level and two rights-based organisations, and included representatives from NEMA, the NCC, Nairobi Water and Sewerage Company (NAWASCO), Umande Trust and Sanergy.

The revealed institutional arrangements governing sanitation provision included the following:

i.   NEMA enforces the law for the protection and conservation of the environment guided by EMCA-1999. FSM is guided by regulations outlined in the Environment and Coordination (Waste Management), meant to streamline the handling, transportation and disposal of various types of waste. Under the Regulations, NEMA also gives licenses operators who use waste transport vehicles for faecal sludge management.

ii.   NCC has two departments related to FSM: i) Department of the Environment that deals with development of physical infrastructure and the protection and conservation of the environment and ii) Public Health Department that enforces the relevant laws and regulations.

iii.   NCC is mandated to provide business permits and enforce public health role.

iv.   NCC has two departments related to FSM: i) Department of the Environment that deals with development of physical infrastructure and the protection and conservation of the environment and ii) Public Health Department that enforces the relevant laws and regulations. NCC is mandated to provide business permits and enforce public health role.

v.   NCWSC is a public utility owned by NCC but managed under the arrangement of the Ministry of Water. It was established under the provisions of the water act 2002. It is charged with the responsibility of providing water and sewerage services to the residents of Nairobi City. NCWSC is mandated to provide faecal sludge disposal points from within the city to either a sewage treatment plant or other designated points.

Interviews with personnel/representatives from NEMA, NCC and NCWSC revealed that:

a. There are overlaps in the institutional set-up of the urban sanitation sector, with regulatory functions being shared by different institutions. As such, proper FSM is fragmented over 3 different departments: i) NEMA provides licenses, ii) NCC provides business permits and enforces public health rules, whereas iii) NCSWC provides disposal points of FS.

b. There is a distinct gap between the demand for and regulated provision of sewerage services, with private sector entities stepping in, particularly in providing emptying services in the unsewered parts of the city. The private sector entities include both formal partners, i.e. 60 licensed exhauster trucks operating in Nairobi, who are registered by NEMA, and informal entities. Informal entities, mainly manual emptiers', are not registered and thrive in LIHDS. Their services are essential in the current FSM chain due to inaccessibility and overcrowding in LIHDS. They, however, use crude, unhygienic and stigmatised tools.

c. There is inadequate funding towards sanitation enhancement, especially related to expansion of the sewer network.

d. There is a severe lack of professionals on on-site sanitation; most professionals in Kenya are trained on centralised sewer systems. As such, there is a gap in development of technologies applicable under different LIHDS conditions in Kenya.

e. Challenges such as illegal status of Kibera, overcrowding, inaccessibility and unplanned nature limits efforts in proving basic amenities within the settlement. However, there are LIHDS upgrading plans by the Government that will give rise to planned high rise apartments equipped with key amenities such as sewer networks, (multitap) water distribution, and sanitation blocks.

f. On-site sanitation has previously been neglected in Kenya, and was mainly considered as a solution for rural setups. A policy framework that will recognise and guide administration, planning, implementation and ownership of on-site sanitation is being developed.

2.3.2   Bio-resource based approach to sanitation

Two agencies in Nairobi, Sanergy and Umande Trust have innovated bio-resource based approach to sanitation that meets the needs of the LIHDS. Their key focus is on improvement of livelihoods in addition to providing safe and hygienic sanitation solutions. In addition to sanitation enhancement, new products have been introduced to the market, as end products of faecal waste processing, namely, proteins (in the form of black soldier fly larvae), organic manure (compost), and fuel (biogas).

17

Sanergy - Application of Urine Diverting Dehydrating Toilets (UDDT)

Sanergy is a social enterprise working on sanitation improvement in informal slum settlements, Nairobi, Kenya (http://www.sanergy.com/pages/about-creative/). Sanergy leverages the entire sanitation value chain to make sanitation provision profitable, and sustainable. With over 1000 Fresh Life Toilets (FLTs), Sanergy collects and transport approximately 5 Kilotonnes of UDDT-FSannually, to its central treatment plant and converts it to safe end products. Sanergy's business model adopts all fundamental aspects in faecal waste management, namely, toilet construction, waste collection & transportation, and waste conversion.

Fresh life Toilet (FLT) construction: Sanergy's sanitation provision approach starts with designing and manufacturing of high-quality, low-cost urine diverting dehydrating toilets (UDDT), referred to as FLTs. Sanergy adopted the franchise model for distribution and operation of the toilets in the community. The model creates economic opportunities for slum residents to earn an income while improving the health and well-being of the community. Local residents within the slum settlements purchase FLT at KES 45,000/= and operate them on pay and use system, charging between KES 5-KES 10 depending on its location within the settlement. The FLT operators, referred to as Fresh Life Operators (FLO's) are franchise partners in an arrangement where Sanergy provides:

i. FLTs, training, ongoing operational and marketing support.
ii. Daily monitoring of sanitation facilities to ensure consistent supply of quality sanitation facilities for customers and boost demand for services.
iii. Land acquisition - Sanergy's government relations team deals with complex land issues encountered by potential FLOs by identifying the relevant authorities and drafting convincing requests seeking approval.
iv. Financial assistance - Sanergy in partnership with local credit facilities assist potential FLOs with interest free loans to purchase FLT. In addition, it's mandatory for new FLOs to attend business training where they learn relevant skills including book keeping, toilet cleaning and customer service.

Waste Collection: Waste generated on daily basis is collected and transported to the central treatment plant located at Kinanie, 40km from the city center. This is done by Sanergy employees, referred as Fresh Life Frontline (FLF) who are well trained and properly equipped. With only narrow, rough and unpaved roads acting as access roads in informal settlements, wheelbarrows and handcarts are used to collect the waste, and delivers the waste to designated transfer station serving a designated section within the settlement. Waste from transfer stations is thereafter transported by truck to the off-site central treatment facility.

Waste conversion: Sanergy focuses on resource recovery. Composting is the main treatment option where UDDT-FS is co-composted with sawdust, and other carbon sources such as coffee husks and effective micro-organisms producing an organic fertiliser, evergrow©. Sanergy is equipped with a laboratory where compost quality is tested regularly to ensure compliance with World Health Organisation standards. Evergrow[©] has established its niche in the Kenyan market, with prices comparable to other organic fertilisers, at KES 35/kg. Sanery is also applying Black Soldier Fly (BSF) technology to convert human waste into resources: protein, in the form of BSF larvae and organic manure.

Umande Trust-Application of Bio-centres

Umande Trust business model provides a viable approach for attaining social objective of inclusive sanitation for underprivileged population. They are founded on the belief that modest resources can help significantly improve access to water, sanitation services and clean energy (http://umande.org/). They combine basic service provision with economic empowerment of the LIHDS communities through;

Financing and construction of Bio-centres- Umande Trust secures funds for Bio-centre construction from donors. Upon securing funding, Umande trust under supervision of their technical team engages trained community technical teams (CTT) to install the Bio-centres. Bio-centre projects serve as multipurpose projects since they provide a range of crucial services to the communities they serve. They are three storey buildings with the following basic requirements: an underground dome shaped digester, bathrooms, kitchen and toilets equally divided into male and female users and additional top-floor space for community socialisation.

*Management of Bio-centres-*The management of the Bio-centres is by registered CBOs or self-help groups who, through a competitive and transparent process, win the selection process. The identified CBO manages the Bio-centre via a pay-and-use system, whereas the community centre is rented on monthly basis. The charges include Kenya Shillings (KES) 5 for toilet use, KES 10 for cold water bath, KES 15 for warm water bath and KES 20 for meal preparation. An economic managerial plan is provided by Umande Trust and shows how the proceeds from the Bio-centres are used: 60% of the money is saved in CBO's bank, which are later divided among the members, 30% of the money caters for operation and maintenance cost and includes cleaning and desludging, whereas 10% is deposited in Umande Trust account and facilitates other related sanitation projects within the settlements.

2.3.3   Funding, technology selection and ownership of the sanitation technologies

Information was elicited from 63 respondents involved in financing and construction of sanitation facilities within the settlement. They included international agencies, government, entrepreneurs, churches and well-wishers through fundraising initiatives at community level. Low community involvement was observed during technology selection (Figure 2.1).

Technology choice is dependent on availability of space and overall project cost, overlooking socio-cultural issues as perceived by low community involvement (Figure 2.2).

The space in which the sanitation projects are implemented is availed from donations by the community, whereas in some cases, it was bought from private land owners.

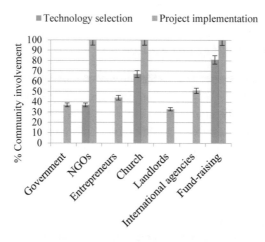

Figure 2.1: Community involvement by sanitation providers during project planning/technology selection and the project implementation phase (n=63)

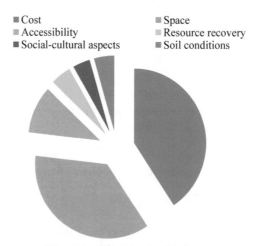

Figure 2.2: Factors considered in technology selection

2.3.4   Operation and Management of the sanitation facilities

A total of 600 sanitation facilities visited were categorised in 3 usage modes: 241 pay-and-use (P&U), 119 free-to-use (FTU), and 240 free & controlled (F&C). Data on financing of the facilities was elicited from the caretakers (Table 2.1).

Table 2.1: Usage mode of sanitation facilities in Kibera.

Financing	Number of facilities	Usage mode		
		P&U	F&C	FTU
Government	71	24	0	47
NGO	103	88	15	
Entrepreneurs	98	98	0	0
Church	8	0	8	0
Landlord	217	0	217	0
International agencies	65	31	0	34
Fundraising	38	0	0	38

A total of 600 sanitation facilities were visited which comprised of four different technology types, of which its frequency distribution is shown in Figure 2.3. The frequency distribution does not reflect the actual users of the facility. It should be noted that bio-digester facilities are visited by about 500-800 people per day. Figure 2.4 shows the condition of the sanitation facility at the moment of visiting during our assessment, termed "general spot check status".

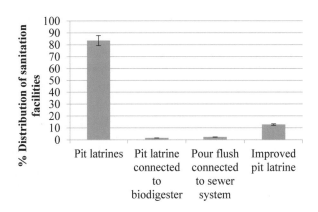

Figure 2.3: Distribution by type of sanitation facilities in the visited areas of Kibera LIHDS

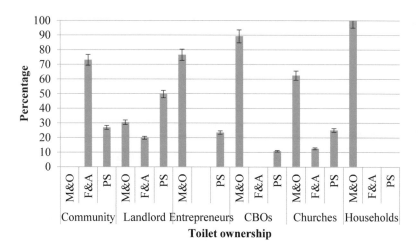

Figure 2.4: General spot check status /condition of sanitation facilities visited during our assessment. Legend: M&O-Maintained and open, F&A-Fill and abandoned and PS-Poor state

The general observed status of the facilities was categorised into: i) Maintained and Open (M&O)-percentage of toilets that were maintained, open and well operated and, ii) Fill and abandon (F&A)-Percentage of toilets that were abandoned after filling up, iii) Poor state (PS)- Percentage of toilets that were in poor state i.e. not cleaned.

Community FTU toilets are left unattended thus; operation and maintenance data were not available. During the assessment, 73% of FTU visited were abandoned after filling up; the 27% still in use were in a poor state. Among the sanitation facilities visited, 36% were financed and implemented by landlords in their individual plots (Table 2.1). 50% of landlord-owned facilities were in poor state, 20% abandoned after filling up and 30% were well managed (Figure 2.4). It was observed that landlords take up the maintenance and operation (M&O) of the sanitation facilities only when tenants pay an additional monthly amount. The additional payment of KES 200 provides a monthly ticket of using the facility for the entire household. Most landlords do not live within the settlement (80%).

The entrepreneur-funded facilities are managed by the owners and a fee is charged to the users in the range of KES 5-KES 10 per use, comparable to the CBO users KES 3-KES 10). Household facilities seemed to provide a good model for sanitation provision. The facilities are owned by the individual households and are available for use by the household members. During our visit, all facilities visited were well maintained. Church-based facilities are owned by churches and are available for use by church members during times of worship, and thereafter remain closed. M&O operations in church-based facilities is carried out using the financial contribution by church members.

Planning of daily O&M of entrepreneur- and CBO-owned facilities is done by sanitation workers, employed by the individual facility owner.

2.3.5   Toilet emptying, transportation and disposal of faecal sludge

Information on pit emptying was elicited from questionnaires administered to 80 pit emptiers within the settlement. The prevalent pit emptying approach within the settlement is manual (96%) with 79% of the tasks being illegal (Figure 2.5). Pit emptying operations are predominantly conducted using buckets (93%), and transportation achieved using hand carts, exhausters or manually using buckets (Figure 2.6). Due to lack of designated faecal sludge disposal points, 85% of the waste ends up untreated in the river. FS disposal within the settlement poses a huge challenge; 4% of the emptiers using exhausters had access to the city's conventional treatment plant, 11% had access to a designated manhole along the main sewer line, whereas 85% of the waste ended untreated in the environment, either via direct disposal to the river or disposal to the storm water drain. NCWSC has not provided any facility for handling the sludge produced from the onsite sanitation facilities. Legal manual pit emptying operations are limited by the higher operational cost (Table 2.2), in comparison to the illegal manual pit emptiers. Although having the lowest desludging cost, illegal pit emptiers use crude, unhygienic and stigmatised methods for FS emptying (Figure 2.8). As such, they mainly work under the influence of drugs. The main challenges which are faced by the pit emptiers are shown in Figure 2.7. Unwillingness to pay for the emptying services lead to F&A strategy, as was seen in the assessed FTU sanitation facilities (Figure 2.4).

■ Vacuum trucks   ■ Manual Legal   ■ Manual Illegal

Figure 2.5: Pit emptying operations in Kibera LIHDS

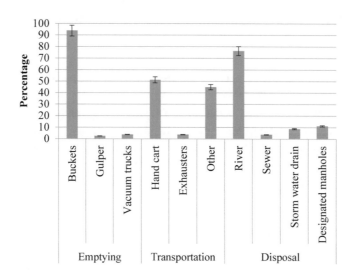

Figure 2.6: Emptying, transportation and disposal mode of faecal waste in Kibera LIHDS

Table 2.2: Costs in Kenyan shilling charged by desludgers for latrine emptying services within settlement; data obtained from pit emptiers

	Liters	Charges (KES)	Charge per liter of sludge (KES)
Vacuum trucks	4000	5000	1.25
Manual-legal	240	500	2.08
Manual-illegal	240	200-300	0.8-1.25

■ Unwillingness of the people to pay
■ Poor accessibility
■ Lack of disposal site

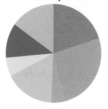

Figure 2.7: Categorised views of pit emptiers regarding their main challenges at work

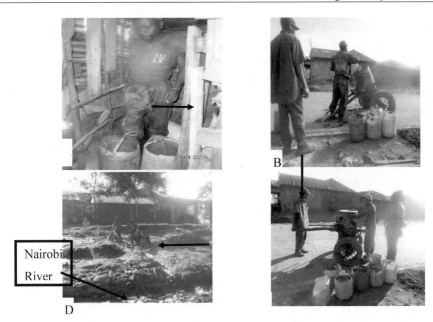

Figure 2.8: Pictures illustrating faecal sludge management activities A) Latrine desludging, B) and C) Preparation for transportation, D) Dumping in Nairobi river

2.4 Discussion

2.4.1 Sanitation provision approach

In sanitation planning, ideally an attempt is made to identify the appropriate sanitation technology for a community, followed by design and implementation (Kalbermatten *et al.*, 1982). This process involves a thorough review of all available technologies for improving sanitation under a given context. Decision-making support tools have been developed to facilitate this process (Loetscher, 1999; Lüthi *et al.*, 2011; Murphy *et al.*, 2009; Tilley *et al.*, 2008). For a proper assessment, sanitation interventions should focus on the users as key stake holder, rather than the suppliers of the interventions, so that the local demand for sanitation services will trigger the supply, operation and maintenance of sanitation systems (Murray & Ray, 2010). In many occasions during sanitation planning in Kibera, the users where not seriously consulted and project cost and availability of space were regarded as the key considerations, whereas accessibility, social cultural concerns and soil condition received low priority (Figure 2.2).

Active community involvement in technology selection was observed in 37%, 67% and 81% of the NGO, church and community owned facilities, respectively (Figure 2.1) and none involvement at all among landlords and entrepreneurs owned facilities. However, spot check of

25

sanitation facilities showed that 73% (Figure 2.4) of FTU (community-based) facilities were abandoned. This was probably due to fact that in FTU facilities, no arrangements were made regarding cleaning and emptying operations. In Mumbai India, lack of community involvement in a sanitation project led to one third of the facilities malfunctioning within the first six months and people returned to open defaecation (Nitti & Sarkar, 2003). The situation was reversed by involving the community in all stages of the project and the introduction of a P&U system. In Bwaise III LIHDS, the lack of community involvement led to unsustainable sanitation interventions, where ventilated improved pit latrines (VIPs) were constructed based on their success in other areas with more favourable conditions (Katukiza *et al.*, 2010).

Moreover, considerations of social and cultural behaviour of the community enhance acceptance and thus the sustainability of a project (Putri & Wardiha, 2013). Thus, involving the community in project planning is vital: it increases the sense of ownership of the facilities. Moreover, the community has basic information about the settlement, which can facilitate decision making in selection of technology to be implemented. The main challenge is how sanitation providers and the community can work together in all phases of the project cycle to ensure that the selected technology is socially, economically and environmentally viable and actually leading to the foreseen increased health benefits.

2.4.2 Kibera sanitation chain

Assessment of the sanitation condition in Kibera indicated that implementation of proper sanitation provision in LIHDS should carefully consider: i) construction and installation of sanitation facilities, ii) supply of sanitation products, iii) repair and maintenance of facilities, iv) emptying services, v) transportation, treatment, and safe disposal of waste and, vi) education and/or sensitisation of the community on hygienic practices (Okurut *et al.*, 2015). Each of the above aspects may affect the well-functioning of the entire sanitation service delivery chain, where failure on one may lead to failed attempts to sanitation enhancement. In Kibera, sanitation improvement efforts mainly focused on provision of sanitation facilities. Although NCWSC has been mandated to provide sludge disposal points, it only caters for 15% of produced faecal sludge (Figure 2.5). This low percentage forms a weak link on the entire sanitation service chain, causing a challenge in FS management; in fact, the demand service outstrips the supply. Private sector entities, both formal and informal, especially stepped in to provide pit emptying services within LIHDS. The services are legally regulated by 4% using vacuum trucks and 18% using manual (legal) pit emptiers (Figure 2.5). The low coverage by vacuum trucks (4%) is due low accessibility and competition from manual pit emptiers, who operate at lower cost. Owing to overcrowding, lack of vehicular access, and low potential for investment in the sector, illegal pit emptiers thrive in the settlement (79%, Figure 2.5), with 85% of sludge being disposed in rivers or open water drains (Figure 2.6). The current sanitation condition results in health and environmental problems. In the peri-urban areas of Kampala, spring water sources were found to be contaminated with faecal matter linked pathogens, which was attributed to poor sanitation practices (Katukiza *et al.*, 2010; Kulabako *et al.*, 2007).

Partnership-based sanitation provision improvements provide an entry point for broader initiatives to improve living conditions in informal settlements. For instance, this has been provided via the Umande Trust and Sanergy business model, both registered by NEMA and NCC, which combines basic service provision with economic empowerment of the communities. Umande Trust enlists services of CBOs in the management of Bio-centres, who thereafter contract registered pit emptiers to offer legal emptying services. Similarly, Sanergy entrust services of the community, herein referred to as Fresh Life Operators to manage the toilets on a pay and use strategy. Sanergy takes up collection, transportation and conversion/treatment of UDDT-FS, and the subsequent sale of recovered end products. In Accra, Ghana, a partnership between government and private sector enhanced faecal sludge management (Boot and Scott, 2008). A similar case was reported in Bamako, Mali (Marc *et al.*, 2004). Our assessment study in Kibera showed low partnership between pit emptiers and local authority: 79% (Figure 2.5) of pit emptying operations are illegal. Moreover, fear of arrest is a challenge among the pit emptiers (Figure 2.7). Due to this, FS mostly ends untreated in the environment via unhygienic disposal pathways (Figure 2.8). Provision of sanitation facilities without planned faecal sludge management, results in a second degree of open defaecation, where human waste ends up in the environment untreated, not directly from an individual but through a second, waste conveying, entity.

2.4.3 Legal and institutional framework

The assessed cases showed that the institutional set-up and regulatory functions of the urban sanitation sector are shared by different institutions. Whereas in Nairobi, the sewerage services fall under the authority of the Ministry of Water, onsite sanitation is under the Ministry of Health. Moreover, NCWSC is mandated to provide FS disposal points, NCC is mandated to issue business permits for sanitation business and enforce public health rules, while NEMA regulates discharge into the sewerage system (Mansour *et al.,* 2017). There is as such need to strengthen the existing legal and institutional frameworks to clarify roles, encourage private sector participation, civil society, and community participation in the sanitation chain.

Our study supports a recommendation that seeks to establish an institutional framework by the government that supports formation of an umbrella body that strengthens the enabling environment between government and all stakeholders of sanitation services (Boot & Scott, 2008). The umbrella body would facilitate exchange of ideas, sharing information and harmonising databases on LIHDS sanitation that shows the sanitation condition at a glance and points out areas requiring improvements. It also creates an avenue where the stakeholders discuss about the challenges and constraints experienced. Thus, paving way to a more sustainable solution, whereas coordination of the activities opens doors for a dialogue, whereby the providers identify the most sustainable technologies in LIHDS.

2.5 Conclusions

This study sought to evaluate factors limiting sanitation provision in LIHDS. The results show that first of all, there is a need for a policy framework to guide planning, implementation and management of onsite sanitation at city level. This would enhance coordination activities of various stakeholders (involved in LIHDS sanitation), ensuring all activities relating to FS are regulated. Community involvement is in itself not adequate to enhance sanitation through community-owned sanitation facilities: they are left abandoned on fill-up. However, a pay-and-use approach of sanitation provision enhances operation and maintenance initiatives: whereas in 73% free-to-use facilities were abandoned on fill-up, 89% and 77% of community-based organisation (CBOs) and entrepreneur-managed facilities, respectively, were well managed.

With, 98% of sanitation facilities in Kibera LIHDS being pit latrines, serving approximately 1 million residents, capacity building programs on non-sewered sanitation are absolutely required. Key aspects under the program would address the aspects of research, development, testing and implementation (RDTI) of alternative options. This would ensure that a wide range of technologies become available for agencies/organisations working on LIHDS sanitation enhancement. The RDTI approach formed the basis for our research, with thesis chapters 3, 4 and 5, exploring off-site co-digestion of faecal sludge with mixed market waste as a means to sanitise excreta for agricultural reuse. Concomitantly, biochemical energy recovery in the form of biogas from FS in LIHDS is pursued as a potential treatment technology.

Partnership-based sanitation provision improvements provide an entry point for broader initiatives to improve living conditions in informal settlements. By involving government actors, CBOs and community, Umande Trust and Sanergy have created an economically viable approach of inclusive sustainable sanitation for underprivileged population. Results from the study can be used by the local government and other partners involved in sanitation improvement within the settlement, to table the importance of a sanitation chain approach and to focus on management schemes that showed to be most sustainable.

Acknowledgements

This research is funded by the Bill & Melinda Gates Foundation under the framework of SaniUp project (Stimulating local Innovation on Sanitation for the Urban Poor in Sub-Saharan Africa and South-East Asia) (OPP1029019). The authors would like to thank Ani Vallabhaneni, Sanergy Kenya, and DVC-ARS, Meru University, Kenya for their valuable support during this study.

References

AWF. 2013. Expanding branded toilet entrepreneurship for improved sustainable sanitation in poor neighbourhoods of Nairobi, Kenya.

Boot, N., Scott, R. 2008. Faecal sludge management in Accra, Ghana: strengthening links in the chain: 33rd WEDC International Conference, Accra, Ghana, 2008. http://wedc.lboro.ac.uk/resources/conference/33/Boot_NLD.pdf.

Evans, B., Tremolet, S. 2010. Output-based Aid and Sustainable Sanitation.

Gulyani, S., Talukdar, D. 2008. Slum real estate: the low-quality high-price puzzle in Nairobi's slum rental market and its implications for theory and practice. World Development 36(10), 1916-1937.

Hogrewe, W., D., J.S., A., P.E. 1993. The unique challenges of improving peri-urban sanitation http://pdf.usaid.gov/pdf_docs/pnabp615.pdf. WASH Technical Report No. 86.

Huchzermeyer, M., Karam, A. 2006. Informal Settlements: A Perpetual Challenge? S.A, Juta and Company Ltd.

K'Akumu, O.A., Olima, W.H.A. 2007. "The Dynamics and Implications of Residential Segregation in Nairobi." Habitat International 31: 87–99.

Kalbermatten, J.M., Julius, D., Gunnerson, C.G., Mara, D.D. 1982. Appropriate Sanitation Alternatives; A Planning and Design Manual. Published for The World Bank The Johns Hopkins University Press Baltimore and London 1982.

Katukiza, A.Y., Ronteltap, M., Oleja, A., Niwagaba, C.B., Kansiime, F., Lens, P.N.L. 2010. Selection of sustainable sanitation technologies for urban slums — A case of Bwaise III in Kampala, Uganda. *Science of The Total Environment*, 409(1), 52-62.

KESSF. 2016. Kenya Environmental Sanitation and Hygiene Strategic Framework (KESSF). *https://www.wsp.org/sites/wsp.org/files/publications/Kenya%20Environmental%20Sa nitation%20 and%20Hygiene%20Strategic%20Framework.pdf.*

KNBS. 2009. Kenya Population and Housing Census.

Kulabako, N.R., Nalubega, M., Thunvik, R. 2007. Study of the impact of land use and hydrogeological settings on the shallow groundwater quality in a peri-urban area of Kampala, Uganda. *Science of The Total Environment*, 381(1–3), 180-199.

Loetscher, T. 1999. Appropriate Sanitation in Developing Countries, Queensland, Brisbane, Australia.

Lüthi, C., J. , McConville, A., Norström, A., Panesar, R., Ingle, D., Saywell, T., Schütze. 2009. Rethinking Sustainable Sanitation for the Urban Environment. In, edited by, pp. The 4th International Conference of the International Forum on Urbanism (IFoU) 2009 Amsterdam/Delft

Lüthi, C., Morel, A., Tilley, E., Lukas Ulrich, L. 2011. Community led urban environmental sanitation planning: CLUES. http://www.wsscc.org/resources/resource-publications/community-led-urban-environmental-sanitation-planning-clues.

Mansour, G., Oyaya, C., Owor, M. 2017. Water and for the urban poor: Sanitation and Situation analysis of the urban sanitation sector in Kenya.

Mara, D., Lane, J., Scott, B., Trouba, D. 2010. Sanitation and health. PLoS medicine 7(11): 590 e1000363.

Marc, J., Doulaye, K., Martin, S. 2004. Private sector management of faecal sludge: A model for the future? Focus on an innovative planning experience in Bamako, Mali. Department of Water and Sanitation in Developing Countries (SANDEC), Switzerland.

MOH. 2016. Kenya Environmental Sanitation and Hygiene Policy 2016-2030.

Mosler, H.J. 2012. A systematic approach to behavior change interventions for the water and sanitation sector in developing countries: a conceptual model, a review, and a guideline. *Int J Environ Health Res*, 22(5), 431-49.

Murphy, H.M., McBean, E.A., Farahbakhsh, K. 2009. Appropriate technology – A comprehensive approach for water and sanitation in the developing world. *Technology in Society*, 31(2), 158-167.

Murray, A., Ray, I. 2010. Commentary: Back-End Users: The Unrecognized Stakeholders in Demand-Driven Sanitation. *Journal of Planning Education and Research*, 30(1), 94-102.

NCWSC/AWSB. 2009. Strategic guidelines for improving water and sanitation services in Nairobi informal settlements. http://www.wsp.org/sites/wsp.org/files/publications/Af_Nairobi_Strategic_Guidelines.pdf.

Nitti, R., Sarkar, S. 2003. Reaching the Poor through Sustainable Partnerships: The Slum Sanitation Program in Mumbai, India. The World Bank, Washington, DC.

Okurut, K., Kulabako, R.N., Chenoweth, J., Charles, K. 2015. Assessing demand for improved sustainable sanitation in low-income informal settlements of urban areas: a critical review. *Int J Environ Health Res*, 25(1), 81-95.

Peal, A., Evans, B., van der Voorden, C. 2010. Hygiene and Sanitation Software: An Overview of 624 Approaches.

Putri, P.S.A., Wardiha, M.W. 2013. Identification problems in the implementation plan of appropriate technology for water and sanitation using FGD approach (case study: Kampong Sodana, Sumba Island, East Nusa Tenggara Province). *Procedia Environmental Sciences* (17), 984 - 991.

Saleem, M., Burdett, T., Heaslip, V. 2019. Health and social impacts of open defecation on women: a systematic review. *BMC public health*, 19(1), 158-158.

Schouten, M.A.C., Mathenge, R.W. 2010. Communal sanitation alternatives for slums: A case study of Kibera, Kenya. *Physics and Chemistry of the Earth, Parts A/B/C*, 35(13–14), 815-822.

Strande, L., Ronteltap, M., Brdjanovic, D. 2014. Faecal Sludge Management Systems Approach for Implementation and Operation. IWA publishing, ww.iwapublishing.com.

Syagga, P. 2011. "Land Tenure in Slum Upgrading Projects." Les cahiers d'Afrique de l'est: 103-113.

Tilley, E., Luthi, C., Morel, A., Zurbrugg, C., Schrtenleib, R. 2008. Compendium of sanitation systems andtechnologies.

http://www.watersanitationhygiene.org/References/EH_KEY_REFERENCES/SANIT ATION/Latrine%20Design%20and%20Construction/Compendium%20of%20Sanitati on%20Systems%20and %20Technologies%20(EAWAG).pdf.

Troeger, C., Forouzanfar, M., Rao, P.C., Khalil, I., Brown, A., Reiner, R.C., Fullman, N., Thompson, R.L., Abajobir, A., Ahmed, M., Alemayohu, M.A., Alvis-Guzman, N., Amare, A.T., Antonio, C.A., Asayesh, H., Avokpaho, E., Awasthi, A., Bacha, U., Barac, A., Betsue, B.D., Beyene, A.S., Boneya, D.J., Malta, D.C., Dandona, L., Dandona, R., Dubey, M., Eshrati, B., Fitchett, J.R.A., Gebrehiwot, T.T., Hailu, G.B., Horino, M., Hotez, P.J., Jibat, T., Jonas, J.B., Kasaeian, A., Kissoon, N., Kotloff, K., Koyanagi, A., Kumar, G.A., Rai, R.K., Lal, A., El Razek, H.M.A., Mengistie, M.A., Moe, C., Patton, G., Platts-Mills, J.A., Qorbani, M., Ram, U., Roba, H.S., Sanabria, J., Sartorius, B., Sawhney, M., Shigematsu, M., Sreeramareddy, C., Swaminathan, S., Tedla, B.A., Jagiellonian, R.T.-M., Ukwaja, K., Werdecker, A., Widdowson, M.-A., Yonemoto, N., El Sayed Zaki, M., Lim, S.S., Naghavi, M., Vos, T., Hay, S.I., Murray, C.J.L., Mokdad, A.H. 2017. Estimates of global, regional, and national morbidity, mortality, and aetiologies of diarrhoeal diseases: a systematic analysis for the Global Burden of Disease Study 2015. *The Lancet Infectious Diseases*, 17(9), 909-948.

Tumwebaze, I.K., Orach, C.G., Niwagaba, C., Luthi, C., Mosler, H.J. 2013. Sanitation facilities in Kampala slums, Uganda: users' satisfaction and determinant factors. *Int J Environ Health Res*, 23(3), 191-204.

UN-Habitat. 2006. Nairobi urban sector profile; Rapid Urban Sector Profiling for Sustainability (RUSPS).

UN-SDG. 2015. The 17 sustainable development goals to transform our world; https://www.un.org/sustainabledevelopment/.

Van Der Geest, S. 2002. The night-soil collector: Bucket latrines in Ghana. Postcolonial Studies: 660 Culture, Politics, Economy 5(2): 197-206.

Van der Hoek, W., Evans, B., Bjerre, J., Calopietro, M., Konradsen, F. 2010. Measuring progress in sanitation.

Wamuchiru, E.K. 2015. Social innovation: the bio-centre approach to water and sanitation infrastructure provision in Nairobi's informal settlement, Kenya. RC21 International Conference on "The Ideal City: between myth and reality. Representations, policies, contradictions and challenges for tomorrow's urban life" Urbino (Italy) 27-29 August 2015. http://www.rc21.org/en/conferences/urbino2015/.

Wasao, S. 2002. Characteristics of Households and Respondents in Population and Health Dynamics in Nairobi's Informal Settlements. African Population & Health Research Centre, Nairobi, Kenya.

Winter, S., Dzombo, M.N., Barchi, F. 2019. Exploring the complex relationship between women's sanitation practices and household diarrhea in the slums of Nairobi: a cross-sectional study. *BMC Infectious Diseases*, 19(1), 242.

Chapter 3: Build-up and impact of volatile fatty acids on *E. coli* and *A. lumbricoides* during co-digestion of urine diverting dehydrating toilet (UDDT-FS) Faeces

This Chapter is based on a paper: Riungu J., Ronteltap, M., van Lier, J.B. 2018. Build-up and impact of volatile fatty acids on *E. coli* an *A. lumbricoides* during co-digestion of urine diverting dehydrating toilet (UDDT-FS) faeces. *J Environ Manage*, 215, 22-31.

Abstract

This study examined the potential of *Escherichia coli* (*E. coli*) and *Ascaris lumbricoides* (*A. lumbricoides*) eggs inactivation in faecal matter coming from urine diverting dehydrating toilets (UDDT-FS) by applying high concentrations of volatile fatty acids (VFAs) during anaerobic digestion. The impact of individual VFAs on *E. coli* and *A. lumbricoides* eggs inactivation in UDDT-FS was assessed by applying various concentrations of store-bought acetate, propionate and butyrate. High VFA concentrations were also obtained by performing co-digestion of UDDT-FS with organic market waste (OMW) using various mixing ratios. All experiments were performed under anaerobic conditions in laboratory scale batch assays at 35±1°C.

A correlation was observed between *E. coli* log inactivation and VFA concentration. Store bought VFA spiked UDDT-FS substrates achieved *E. coli* inactivation up to 4.7 log units/ day compared to UDDT-FS control sample that achieved 0.6 log units/ day. In co-digesting UDDT-FS and organic market waste (OMW), a ND-VFA concentration of 4800-6000 mg/L was needed to achieve *E. coli* log inactivation to below detectable levels and complete *A. lumbricoides* egg inactivation in less than four days. *E. coli* and *A. lumbricoides* egg inactivation was found to be related to the concentration of non-dissociated VFA (ND-VFA), increasing with an increase in the OMW fraction in the feed substrate. Highest ND-VFA concentration of 6500 mg/L was obtained at a UDDT-FS:OMW ratio 1:1, below which there was a decline, attributed to product inhibition of acidogenic bacteria. Results of our present research showed the potential for *E. coli* and *A. lumbricoides* inactivation from UDDT-FS up to WHO standards by allowing VFA build-up during anaerobic digestion of faecal matter

3.1 Introduction

As an innovative solution for enhancing sanitation in low income urban areas, urine diverting dehydrating toilets (UDDTs) can be offered on a pay-and-use basis in the form of serviced shared facilities. The UDDT principle involves separate collection of faeces and urine (Austin, 2001; Austin & Cloete, 2008; Niwagaba *et al.*, 2009a; Sherpa *et al.*, 2009). The above is the set-up of Sanergy, Kenya, a company working on sanitation improvement within Mukuru Kwa Njenga and Mukuru Kwa Reuben LIHDS, Kenya. After every use, sawdust is sprinkled on separated faeces mainly for odour and flies elimination (Austin & Cloete, 2008; Niwagaba *et al.*, 2009a). However, addition of saw dust or ash is not sufficient to kill pathogens (Niwagaba *et al.*, 2009a). Thus, an extra pathogen inactivation step is required after waste collection, especially when the faecal matter will be valorised for agricultural purposes. Some of the treatment options associated with source separated human waste include anaerobic digestion, composting, ash addition, chemical treatment and storage (Fagbohungbe *et al.*, 2015; Larsen & Maurer, 2011; Niwagaba *et al.*, 2009b; Rajagopal *et al.*, 2013; Vinnerås, 2007).

Anaerobic digestion (AD) offers an attractive approach in human waste treatment (Rajagopal *et al.*, 2013). It provides organic waste treatment by avoiding volatile organic compounds emissions, digestion of organic matter, build-up of an effluent with good fertilising qualities in addition to energy recovery through methane build-up (Avery *et al.*, 2014; Fonoll *et al.*, 2015; Nallathambi Gunaseelan, 1997; Romero-Güiza *et al.*, 2014). However, studies have reported unsatisfactory pathogen inactivation in AD (Chaggu, 2004; Dudley *et al.*, 1980; Foliguet & Doncoeur, 1972; Leclerc & Brouzes, 1973; McKinney *et al.*, 1958; Pramer *et al.*, 1950). Thus, sludge produced requires a post-treatment step, which can be expensive, time consuming, or may create pathways for disease transmission. As such, more research is needed on enhancing pathogen inactivation during the overall anaerobic digestion process.

Some key factors influencing pathogen inactivation during AD include: temperature and time (Gibbs *et al.*, 1995; Olsen *et al.*, 1985; Olsen & Larsen, 1987), reactor configuration (Kearney *et al.*, 1993; Olsen *et al.*, 1985), pH and VFA concentration (Abdul & Lloyd, 1985; Farrah & Bitton, 1983; Sahlström, 2003).

Basically, AD is a biological process where organic matter (carbohydrates, lipids, proteins) except lignin components, is degraded in the absence of oxygen, producing methane and carbon dioxide (Jankowska *et al.*, 2015; Van Lier *et al.*, 2008), with main processes involved being hydrolysis, acetogenesis, acidogenesis and methanogenesis. During the treatment of solids-rich waste streams, hydrolysis is the rate determining step (Van Lier *et al.*, 2008) with pH, temperature, C/N ratio and hydraulic retention time (HRT) being reported as the key factors controlling the VFA build-up (Chen *et al.*, 2007; Lee *et al.*, 2014; Wang *et al.*, 2014a), and its subsequent conversion into methane.

VFAs are commonly produced during the hydrolysis/ acidogenesis stage of anaerobic digestion. This stage has widely received attention within various studies focusing on either enhancing methane build-up, bio-hydrogen generation, pathogen inactivation or VFA production (Battimelli *et al.*, 2009; Ghimire *et al.*, 2015; Ghosh *et al.*, 1985; Kim *et al.*, 2011; Mata-Alvarez, 1987; Palmowski *et al.*, 2006). In addition, hydrolysis increases solubilisation of the particulate organic fraction in the feed mixture (Rajagopal *et al.*, 2013).

The degree of disinfection achieved in anaerobic digestion is influenced by a variety of interacting operational variables and conditions (Smith *et al.*, 2005). Reactor configuration, hydraulic retention time (HRT), organic loading rate (OLR), temperature, VFA and pH have all emerged as critical variables affecting pathogen inactivation. Bacteria inactivation due to temperature is related to time, with digestion under thermophilic conditions requiring less time for bacterial inactivation than under mesophilic conditions. *M. paratuberculosis* and *Salmonella* were inactivated within 24 hours under thermophilic conditions compared to months under mesophilic anaerobic digestion (Olsen *et al..*, 1985).

In addition, pH and VFA concentrations in the feed substrate may determine bacterial survival during anaerobic digestion (Abdul & Lloyd, 1985; Farrah & Bitton, 1983; Sahlström, 2003). VFA toxicity is associated with the dissociation of the acid molecule: non-dissociated VFAs are able to pass through the cell membrane of microbes by passive diffusion and will dissociate internally, disturbing internal pH, impacting protein's tertiary structure, and inhibiting microbial growth (Jiang *et al.*, 2013; Wang *et al.*, 2014a; Zhang *et al.*, 2005). Additionally, non-dissociated (ND)-VFAs can make the cell membrane permeable, which allows leaching of the cell content and disintegration of the microbes. The antibacterial effects of ND-VFA have been demonstrated in treatment of enteric *E. coli* infections of rabbits and pigs, where a rise in caecal pH in diarrhoeic condition over the normal pH was cited as the main infection cause (Prohászka, 1980a; Prohászka, 1986a): at higher pH, less ND-VFA was present to inactivate pathogens.

As much as this technique is promising, adjusting reactor settings to maximise the ND-VFA fraction is not yet a standard practice; more research is needed in order to understand the possible sanitising effect of VFAs during the anaerobic fermentation of source separated human faeces.

As such, this study assessed the potential of pathogen inactivation by VFAs in an acidogenic reactor. The research is part of a bigger research project investigating ways to enhance pathogen inactivation and biogas production from faecal matter coming from urine diverting dehydrating toilets (UDDT-FS), at Sanergy Kenya. Provided in this paper are laboratory batch scale results of:

i.   Comparing *E .coli* inactivation in VFA spiked and non-spiked UDDT waste samples
ii.   Assessing effect of initial pH and UDDT-FS:OMW mix ratio on VFA build-up and pathogen inactivation.

3.2 Materials and Methods

3.2.1 UDDT-FS waste samples

UDDT-FS samples used for this study were obtained from the Fresh Life® urine diverting dry toilets (UDDT) within Mukuru Kwa Njenga/ Mukuru Kwa Reuben LIHDS, Kenya. The Fresh Life® toilets are fabricated and installed by a social enterprise, Sanergy, in collaboration with entrepreneurs in the informal settlements who maintain them. Within each toilet facility, a 30 l container is used for waste collection, with approximately 10 g sawdust added after every toilet use. The toilets are emptied on daily basis, where used containers are replaced with clean ones. Every day, approximately 7 tons of UDDT-FS are collected and transported to an offsite central treatment location (in Machakos County Government approximately 30 km from the city of Nairobi).

Five containers with UDDT-FS were randomly selected after which mixing of the contents was done in order to obtain a homogeneous mix. From each of the ten containers, 1 kg UDDT-FS was sampled and transferred into a plastic container. Further mixing of this waste was done in order to homogenise the sample.

3.2.2 Organic market waste samples

OMW was collected from vegetable vendors, eating points and waste disposal points within Mukuru Kwa Njenga and Mukuru Kwa Reuben LIHDS. 15 kg of the waste was collected and contained food waste, vegetable waste and fruit waste, in equal proportions. After collection of the waste, 2 kg of each waste was sampled out and further separately homogenised by use of a domestic blender for one minute.

In readiness for the experiments, the waste was then mixed in three different UDDT-FS:OMW ratios (by weight): 4:1, 1:1 and 1:0, and refrigerated at 4°. Table 3.1 shows the characteristics of UDDT-FS and OMW waste used in the study.

Table 3.1: Characterisation of urine diverting dehydrating toilets waste and mixed organic market waste used in study

	UDDT-FS		OMW	
	Value	STDEV	Value	STDEV
TS (% wgt)	24.5	3.8	17.9	1.6
Moisture content	75.5	3.8	80.7	4.1
VS (% wgt)	20.1	3.5	16.9	4.4
***E. coli* (CFU/g TS)**	1.7E+09	5.3E+08	2.7E+05	7.4E+04
Ascaris eggs	Not detected		Not detected	

3.2.3 Fertilised *A. lumbricoides* eggs

These were purchased from Mahidol University School of Tropical Diseases, Bangkok. They were packed in 50 ml tubes, each with 100,000 eggs suspended in 0.1% formalin. A viability test conducted on the eggs showed that they were 100% viable. Eggs were stored at 4°C before use.

3.2.4 Laboratory Scale Experimental set-up

This section describes the methodology used in the three different experiments. Other than the substrate preparation, same test conditions applied. Waste collection and preparation is described in section 2.1.2.

The experiments were conducted in 0.1 L serum bottles, maintaining a working volume of 80 ml. After substrate addition to the bottles, they were tightly closed with rubber stoppers and flushed with argon gas in order to create anaerobic conditions (Jensen *et al.*, 2011). The bottles were then incubated at 35±1°C. Mixing of the serum bottles was done manually.

E. coli inactivation in VFA spiked versus VFA un-spiked waste samples

VFA concentration effect was tested by separately spiking serum bottles containing UDDT-FS substrate, 15 g TS/L concentration with store bought acetic, propionic and butyric acids (supplied by Sigma Aldrich, (ACS reagent, ≥99%)), each at three different concentrations; 2000, 3000 and 4000 mg/L, and adjusting the pH of the serum bottle contents to 4.8 ($pH_{4.8}$) with 0.1 M HCl. By adjusting the initial pH to 4.8, which is the average pK_a value of the acids, the concentration of non-dissociated acids and dissociated acids was set to a 1:1 ratio. In addition, three controls were set in: 1) UDDT-FS with a fixed pH at 4.8 ($pH_{4.8}$), and 2) UDDT-FS without pH adjustment (pH_{un}), and 3) organic market waste (OMW) without pH adjustment (pH_{un}). For each of these experiments, a total of three serum bottles was set, so that one was sacrificed each day for analysis of *E. coli*, pH and VFA. The serum bottles were then closed with rubber stoppers and the procedure described in section 2.2 was followed.

UDDT-FS: OMW mix ratio on VFA build-up and pathogen inactivation

Each serum bottle contained on average 4 g VS/100 mL of the waste substrate. Experiments were conducted at two initial pH levels: 4.8 ($pH_{4.8}$), (adjusted using 0.1M HCl or NaOH solution) and unaltered substrate pH (pH_{un}). Table 3.2 shows UDDT-FS:OMW mix ratios and respective pH levels as used in the experiments. For each experiment, five serum bottles were set, three for VFA, *E. coli* and pH and two for *A. lumbricoides* eggs analysis. For VFA, *E. coli* and pH, one was sacrificed on day 1, 3 and 4 for analysis. The two serum bottles set for helminth egg analysis were spiked with fertilised *A. lumbricoides* eggs at a concentration of 20 eggs/ml in order to evaluate efficiency of the treatment process in their inactivation. Extraction of *A. lumbricoides* eggs for viability checks was done on day 2 and 4.

Table 3.2: UDDT-FS: OMW mix ratios and respective pH levels as used in the experiments

pH level	UDDT F: OMW mix ratio	1:0	4:1	2:1	1:1	1:2	1:4	0:1
Adjusted	Initial pH	4.8	4.8	4.8	4.8	4.8	4.8	4.8
	Final pH	5.1	4.4	4	3.7	3.5	3.3	3.1
Unaltered	Initial pH	5.8	5.7	5.7	5.6	5.6	5.5	5.3
	Final	5.4	4.9	4.4	3.8	3.6	3.4	3.2

3.2.5 Analytical procedures

3.2.5.1 Total solids and volatile solids

Total solids (TS) and volatile solids (VS) analysis were conducted according to the gravimetric method (SM-2540D and SM-2540E), as outlined in Standard Methods for the Examination of Water and Wastewater (APHA, 1995a). TS was analysed by drying 10 g sample weight in an oven at 105 °C for 24 hours, after which they were cooled and weighed. For VS determination, the samples were further dried in a muffle furnace at 550°C for 2 hours. pH measuring was done using an analogue pH/ORP meter (model HI8314-S/N 08586318) calibrated with buffer solution at pH 4 and pH 7.

3.2.5.2 VFA measurements

Volatile fatty acids (VFA) in wastewater samples can be measured using gas chromatographical methods for organic acids or by titrimetric method (Jobling Purser *et al.*, 2014; Lützhøft *et al.*, 2014). In this study, the titrimetric method as described in analytical methods for waste water characterisation and evaluation of reactor performance during anaerobic treatment (AGROIWATECH, DeliverableD2) was adopted. The sample, placed in centrifuge tubes was centrifuged for 5 minutes at 5,000 revolutions per minute (rpm). 50 ml of the supernant was put in a beaker. The pH of the solution was adjusted to 6.5 using either 0.1M HCl or NaOH solution, and thereafter it was titrated with 0.1M HCl to pH 3. This volume of acid was recorded. The samples were transferred to a digestion flask with glass beads and connected to a condenser. Flask was heated until liquid began to boil and then allowed for three minutes boiling. The heater was then switched off and two minutes were allowed for cooling. The sample was then titrated immediately to pH 6.5 and volume of base recorded. The total VFA values in meq/ l added were then calculated using the formula (AGROIWATECH, DeliverableD2):

$$\left(\frac{B \times 0.1 \frac{meq}{mL}}{C}\right) \times 1000 \dots \dots (1)$$

Where B is the volume of 0.1 M sodium hydroxide required to titrate from pH 3-6.5; C the total volume of titrated sample (ml), 0.1-meq conversion factor, 1000- ml to l conversion.

In addition, background acidity correction was carried out to correct for proton acceptors present in the waste water that are not volatile, e.g. humic acids. For this, the prescribed procedure was carried out on fresh UDDT-FS and OMW samples and obtained VFA values subtracted from incubated sample values. The procedure also accounted for actual VFAs in the sample before treatment, so as to set a baseline for VFA build-up after treatment.

From the Total Volatile Fatty Acid (TVFA) concentration, the fraction of ND-VFA was calculated. VFAs are commonly considered to constitute a single weak-acid system with equilibrium constant Ka because of the similarity of their pK values (Lahav & Morgan, 2004; Moosbrugger R. E. *et al.*, 1993). Therefore:

$$((H^+) \times (H^-))/(HA) = Ka \dots \dots (2)$$
$$pH = pKa + Log_{10}\left(\frac{A^-}{HA}\right) \dots \dots (3)$$
$$A_T = (HA) + (A^-) \dots \dots (4)$$

Where: AT = total VFA species concentration (mg/L), HA represents the acidic, protonated species and A⁻ the ionised form of each acid.

3.2.5.3 *E. coli* enumeration

E. coli enumeration was done using the chromocult coliform agar (CCA) technique, which was proved applicable for use in temperate regions (Buckley *et al.*, 2008; Byamukama *et al.*, 2000; Frampton *et al.*, 1988; Manafi & Kneifel, 1989). The CCA (Chromocult; Merck, Darmstadt, Germany) was prepared following manufacturer's instructions. Homogenised samples were serially diluted (10^{-1} to 10^{-6}) with the peptone buffered water. For each sample dilution, 0.1 ml was spread on prepared chromocult agar plates in duplicate. The prepared plates were then incubated for 24 hours at a temperature of 36±1 ^{0}C, after which colony counting was facilitated by use of a colony counter (IUL magnifying glass colony counter, IUL, S.A., Barcelona, Spain). The criteria used for identification were able to identify dark blue- to violet-coloured colonies as *E. coli* (Byamukama *et al.*, 2000; Finney *et al.*, 2003). The average numbers of colonies were used to calculate the *E. coli* concentrations in the samples, expressed in CFU/ g 100 ml of the test sample. In the method, the lowest detection limit is 1000 CFU/L.

The first order reaction coefficients for *E. coli* removal were calculated using the Chick-Watson model that expresses the rate of inactivation of micro-organisms by a first order reaction.

$$\ln \left(C_t / C_0 \right) = -kt \quad \ldots \ldots (5)$$

Where:

$C_t = Number\ of\ micro - organisms\ at\ time\ t$
$C_0 = Number\ of\ micro - organisms\ at\ time\ 0$
$k = decay\ constant$
$t = time$

3.2.5.4 *A. lumbricoides* egg recovery

A. lumbricoides egg recovery was performed according to method developed by Moodley *et al..,* (2008) and modified by Pebsworth *et al..,* (2012). Ammonium bicarbonate solution was added to 80 ml of sample in order to wash and dissociate the eggs attached on the particles. It was then passed through a 100 μm sieve onto a 20 μm sieve. Sieve contents were well washed and all material held on 100-μm discarded. Material held on the 20 μm sieve was washed and collected onto autoclaved 15 ml centrifuge tubes, and centrifuged using a bench top centrifuge (EBA 20, Andreas Hettich GmbH &CO. KG, Germany) at 3000 rpm for 5 minutes. Supernatant was discarded and remaining pellets re-suspended in $ZnSO_4$ (specific gravity 1.3), while vortexing until the 14 ml level. The samples were then centrifuged again at 2000 rpm for 5 minutes. Supernatant was then poured through a small 20-μm filter, and washed off into an autoclaved plastic test tube. It was then centrifuged again at 3000 rpm for 5 minutes. The supernatant was discarded and the egg pellets were transferred into a 50 mL Falcon tube, with de-ionised water added to the 45 ml mark. The Falcon tube was covered with parafilm that was pricked (to allow air exchange within the sample) and then incubated at 28±1 °C for 28 days. Regular checks with de-ionised water additions were conducted in order to account for water lost through evaporation.

After the 28 days incubation, the samples were divided into 15 ml centrifuge tubes and centrifuged for 5 min. at 3000 rpm. The supernatant was discarded and the remaining pellet containing eggs was well mixed using a pipette. 1 ml of sample was transferred on a Sedgewick-rafter counting cell (from Wildlife Supply Company[R]). The slide was observed under the microscope (AmScope, California, USA) at a magnification of 10 and 40.

Eggs developed to the larval stage, with motile larvae, were considered viable, while all eggs that stopped under any other developmental stage and eggs that presented some kind of deterioration with no motile larvae inside were considered non-viable.

3.2.6   Data analysis

Data analysis was done using Microsoft Excel software. Data obtained from each batch was first analysed by computing the averages of the three trials conducted per batch. Average values of the three batches were then combined by computing their average values, standard deviations and standard errors. The average values of the three batches were then presented in either Table of Figure form.

3.3 Results and discussion

3.3.1   VFA concentration effect on *E. coli* inactivation

Volatile fatty acids are produced when larger organic molecules are hydrolysed and anaerobically oxidised to carboxylic acids. Depending on the alkalinity of the solution, this will affect the pH value. With a decrease in pH, the fraction of acids present in the non-dissociated form will increase, which are reported to support bacteria inactivation. In order to study the relationship between non-dissociated VFAs and *E.coli* inactivation, batch experiments were conducted with store-bought VFAs (acetate, propionate and butyrate at 2,000, 3,000 and 4,000 mg/L) and spiked in to 15 g/L TS UDDT-FS samples. Acetate, propionate or butyrate VFA concentration of 4000 mg/l achieved *E. coli* inactivation to below detectable levels after two days of incubation.

Increasing the VFA concentration has a direct effect on *E. coli* inactivation (Figure 3.1): for butyrate, the *E. coli* log inactivation increases from 3 to 6 by increasing the spiking concentration from 2,000 to 4,000 mg/L. A similar trend was observed for the acetic acid and propionic acid spiked bottles. Increasing concentration leads to increase in ND-VFA fraction which increases the inhibitory effect.

The VFA chain length apparently had slight impact on *E. coli* inactivation as indicated by the calculated log inactivation rates (log inactivation per ND-VFA (meq/g TS added)). After one day treatment, with acetic, propionic and butyric acids, the achieved log inactivation was 2.15, 2.13 and 2.11, respectively, showing a slight decrease with increasing chain length of the fatty (Figure 3. 2). Related studies investigating the effect of VFA chain length on *Salmonella typhimurium* and *Vibrio cholera* inactivation reported that inactivation efficiency decreased with increasing chain length (Goepfert & Hicks, 1969b; Kunte *et al.,* 2000; Salsali *et al.,* 2006), which agrees with the trend observed in our present study for *E. coli* inactivation. Acetic acid, due to its lower molecular weight, diffuses faster across the bacterial membrane compared to propionic and butyric acid under the same conditions of temperature and pH. Upon passage of ND-VFA's through the cell membrane of microbes they dissociate internally thus disturbing internal pH, impacting protein's tertiary structure, and inhibiting microbial growth (Jiang *et al.,* 2013; Wang *et al.,* 2014a; Zhang *et al.,* 2005), thus inactivating pathogens. Under neutral pH conditions, increased VFA toxicity with increasing chain length on methanogenesis has been described previously (Van Lier *et al.,* 1993). The increased toxicity was ascribed to the presence of a longer a-polar aliphatic tail that more easily interferes with the bacterial or archaeal membrane.

Experimental results applying a set VFA concentration showed that the *E. coli* inactivation rate increased in time during the incubations: higher *E. coli* inactivation was achieved on the second day of treatment (Figure 3.1). The *E. coli* log inactivation achieved at 3000 mg/L butyric acid, after day 1 and day 2 of treatment was 4.2 and 6.0, respectively. The increased inactivation rate coincided with a drop in pH from 4.8-4.3 between day 0 and day 2 and a corresponding increase in ND-VFA concentration from 21.6-33.9 meq/l. The pH drop from 4.8-4.3 increased protonation of the organic acids thus higher ND-VFA fraction in the waste sample with the overall effect being increased inhibitory. In addition, increased incubation time leads to longer exposure time of *E. coli* to toxic effect of the ND-VFA.

The control sample with UDDT-FS also showed a higher log inactivation rate at lower pH: at 4.8, the *E. coli* log inactivation was 1.2 log/d, while at pH 6.2 the inactivation did not exceed 0.6 log/d. The total amount of ND-VFA increased up to 0.5 meq/ g TS added in UDDT-FS, $pH_{4.8}$ and 0.2 meq/ g TS added in UDDT-FS, pH_{un}. In the VFA spiked incubations, a lower pH was achieved at higher initial VFA concentration.

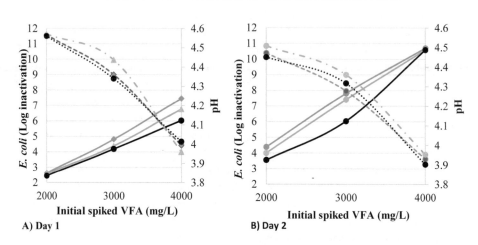

Figure 3.1: pH and *E. coli* inactivation trends at varying initial spiked concentrations of acetate, propionate and butyrate: A) Day 1 of treatment, B) Day 2 of treatment

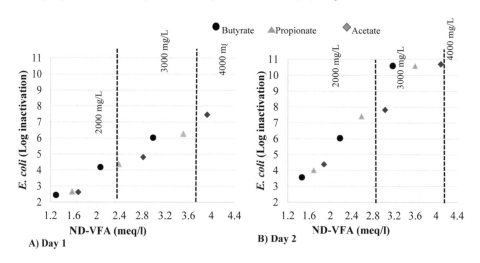

Figure 3.2: Build up and effect of ND-VFA on *E. coli* inactivation at various concentrations of acetate, propionate and butyrate on; A) Day 1 and B) Day 2 of treatment

OMW hydrolysis: The used OMW hydrolysed rapidly, leading to a pH of 3.3 after 24 hours incubation. From an average initial *E. coli* count of 1.75E+07 ±1.3E+07, values were already below detectable levels after the first day. At the measured pH, over 97% of the total VFA was present in non-dissociated form, leading to a ND-VFA concentration of 36.4 meq/l. For OMW, the presence of readily degradable fractions of organics caused a rapid acidification leading to a high concentration of volatile fatty acids and a low pH. The latter is of interest for co-digestion with dried faecal matter since it will accelerate pathogen inactivation, particularly in two-stage or plug flow digestion systems. Further experiments were conducted to assess the potential of OMW co-digestion for UDDT-FS digestion.

3.3.2 Co-digestion of UDDT Waste and Mixed OMW

3.3.2.1 Effect of mix ratio on pH and VFA build-up

The experiments were conducted at various UDDT-FS:OMW mix ratios (1:0 4:1, 2:1, 1:1, 1:2, 1:4 and 0:1) at an average total TS concentration of 40 g/L (Table 3.2). Figure 3.3 shows the ND-VFA and dissociated-VFA (D-VFA) build-up at different UDDT-FS:OMW ratios. Co-digesting UDDT waste with OMW led to a decline in pH, with a stronger decline in samples with a higher OMW fraction (Figure 3.3), which was already observed during the first day of treatment. After incubating the UDDT-FS:OMW ratios of 1:1, 1:2, 1:4 and 0:1, the pH declined within 24 hours to the range 3.1-3.7 for the bottles with an initial adjusted pH of 4.8 and to the range 3.2-3.8 for the unadjusted substrates. Apparently, no pH adjustments are needed for reaching a sufficiently low pH at these ratios. In the experiments with a higher content of UDDT-FS, a more clear difference was observed between the samples with and without adjusted pH. For example, at a UDDT-FS:OMW ratio of 4:1, the final pH was 4.4 and 4.9 in the bottles with initial pH$_{4.8}$ and pH respectively.

Results clearly show that the VFA build-up was dependent on the OMW fraction in the substrate, which subsequently impacted the final pH of the batch incubations. Maximum VFA build-up was achieved at an UDDT-FS:OMW ratio of 1:1, below which a decline was observed both in VFA and ND-VFA build-up, while the pH dropped to below 4.

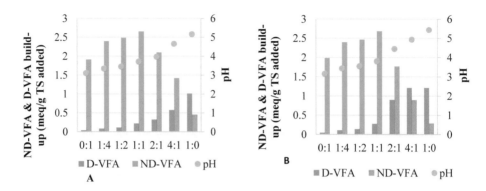

Figure 3.3: ND-VFA and D-VFA build-up in co-digestion experiments at the end of incubation period: a) Initial pH$_{4.8}$, b) Initial pH$_{un}$, at UDDT-FS:OMW ratios 0:1, 1:4, 1:2, 1:1, 2:1, 4:1 and 1:0

Moreover, it was observed that the VFA build-up was not proportional to the amount of OMW added (Figure 3.3). Results show highest ND-VFA build-up (2.7 meq/ g total TS added) at UDDT-FS:OMW=1:1, below which a decline was observed both in D-VFA and ND-VFA. In the latter incubations, the final pH recorded was below 4. VFA toxicity is associated with the degree of dissociation of the carboxylic acid group, which is governed by a pKa of about 4.8. Increasing the OMW fraction in the feed substrate leads to rapid acidification thereby lowering the pH and increasing the ND-VFA concentration. The increased concentration of ND-VFAs has a toxic effect not only on pathogens but also on the anaerobic bacterial population including the VFA producers and methanogenic Archaea. The latter calls for an optimisation of OMW dosing.

3.3.2.2 *E. coli* inactivation in single substrate and co-digestion experiments

UDDT-FS:OMW mix ratio: *E. coli* inactivation increased with an increase in OMW fraction in the feed substrate (Figure 3.4). UDDT-FS:OMW=1:4, 1:2 and 1:1, achieved *E. coli* inactivation to below detectable limits between day 1 and day 2, with corresponding decay rates (k values) in the range of 7.4-13 /day (Figure 3.4), whereas OMW alone showed inactivation to below detectable limits in one day, all meeting WHO standards of <1*10^3 CFU/ 100 ml (WHO, 2006). UDDT-FS:OMW ratio 2:1 showed *E. coli* inactivation to below detectable levels in 3 days, whereas UDDT-FS:OMW ratio 4:1 showed inactivation to 2.0*10^3 ± 1.35*10^3 and 5.6*10^4±6.5*10^5 CFU/100 ml at initial pH 4.8 and unadjusted initial pH respectively. Results clearly showed that *E. coli* decay rates (k value) increased with increase in OMW in the waste substrate (Figure 3.4). Below UDDT-FS:OMW ratio 1:1, an increase in decay rate was observed despite a decline in ND-VFA concentration. As discussed in section

3.2.1, aiming at highest k-values is not considered the best strategy, since increased toxic effect of ND-VFA not only affects pathogens but all anaerobic bacterial population, thus process failure.

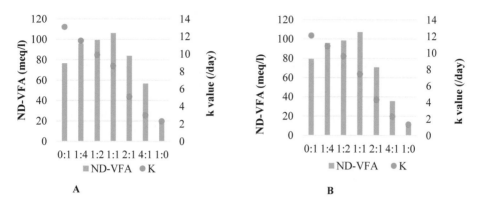

Figure 3.4: ND-VFA concentration and k value trend at UDDT-FS:OMW mix ratios 1:4, 1:2, 1:1, 21, 4:1 and 1:0; a) Initial pH$_{4.8}$, b) Initial pH$_{un}$

It was observed that total ND-VFA concentration in the range of 80-100 meq/l was needed to achieve *E. coli* log inactivation to below detectable levels in four days. Similarly, ND-VFA concentration in the range of 48-72 meq/l caused between 3-5 *E. coli* log inactivation in four days.

pH: Initial substrate pH did not affect *E. coli* inactivation in substrates with high OMW fraction, owing to prevailing VFA production. The case was observed at UDDT-FS:OMW ratios 0:1, 1:4, 1:2 and 1:1, where *E. coli* inactivation to below detectable limits was achieved within the same period of time. However, a reduced *E. coli* inactivation was observed at decreasing OMW fractions in feed substrate, i.e. UDDT-FS:OMW ratios 1:0, 4:1, 2:1. The ratio 2:1, pH$_{4.8}$ achieved *E. coli* inactivation to below detectable limits in four days, whereas at pH$_{un}$, the achieved log inactivation was 5.1 log. At lower OMW fractions, adjusting pH led to an increase in ND-VFA concentrations in the feed substrate.

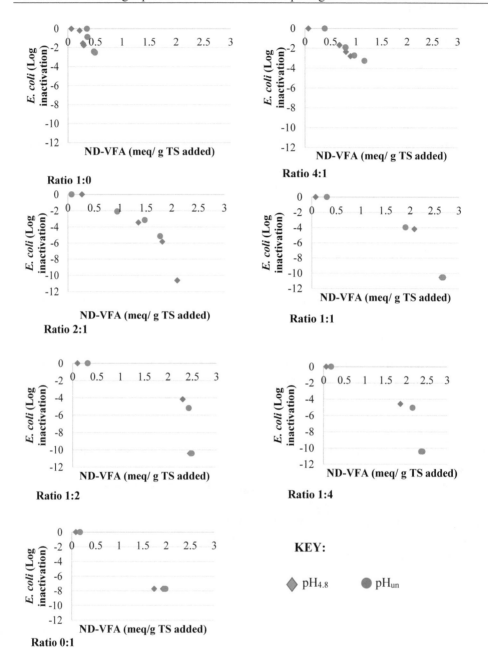

Figure 3.5: ND-VFA build-up (meq/g total TS added) versus *E. coli* log inactivation profiles at various UDDT-FS:OMW mix ratios

3.3.2.3 *A. lumbricoides* egg inactivation

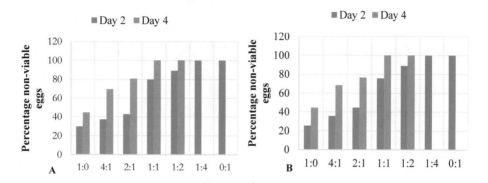

Figure 3.6: *A. lumbricoides* **egg inactivation with time for UDDT-FS:OMW mix ratios 1:4, 1:2, 1:1, 21, 4:1 and 1:0; A. pH 4.8, B. pH$_{un}$**

The higher the OMW fraction in the substrate, the higher was the *A. lumbricoides* egg inactivation achieved. Complete *A. lumbricoides* egg inactivation was achieved within two days in OMW and UDDT-FS:OMW ratio 1:4 substrates at both initial pH values (Figure 3.6). Similarly, complete *A. lumbricoides* eggs inactivation was achieved in OMW and UDDT-FS:OMW ratio 1:2 and 1:1 after four days of treatment, and more than four days are required in UDDT-FS:OMW ratios 2:1, 4:1 and 1:0. UDDT-FS:OMW=4:1 pH 4.8, UDDT-FS:OMW=4:1 pH 5.82, UDDT-FS pH 4.8, UDDT-FS pH 5.74 and UDDT-FS pH 7 recorded low inactivation levels of 73, 70, 65, 68 and 16 % respectively after four days treatment. Overall, our current experimental results showed a rapid inactivation of helminth ova, which depended on the OMW fraction in the feed substrate. Results are in contrast to studies that reported high resistance of helminth ova to inactivation (Jimenez-Cisneros, 2007). The persistence of helminth ova are ascribed to their 3-4 protective layers that increase their resistance to treatment against desiccation, strong acids/ bases, oxidants, reductive agents, detergents and proteolytic compounds (Jimenez-Cisneros, 2007). Figure 3.7 shows microscopic images of *A. lumbricoides* eggs after various treatments in this study.

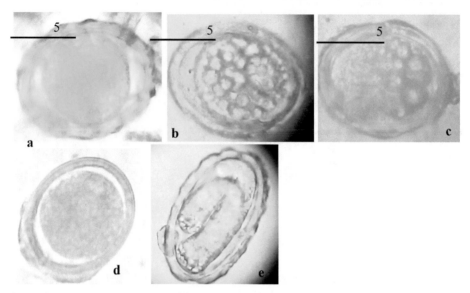

Figure 3.7: Microscopic images of *A. lumbricoides* eggs after two days of incubation at 35°C: a) Fertilised untreated *A. lumbricoides* egg, b) non-viable egg, OMW, pH 4.8; c) Non-viable *A. lumbricoides* egg, UDDT:OMW=1:4, pH 5.48; c) Non-Viable *A. lumbricoides* egg, UDDT-FS, pH 5.7 e) Embroyonated *A. lumbricoides* egg, UDDT-FS pH 7.

3.3.3 Practical application in enhancing UDDT-FS sanitisation

Study results showed increasing *E. coli* inactivation at increasing OMW fraction in the feed substrate until an optimum dosage in a ratio of 1:1.

Results indicate that ND-VFA build-up of 2-2.5 meq/ g total TS was required to achieve *E. coli* inactivation to below detectable level in less than four days. The VFA build-up agrees with a total VFA concentration of approximately 4800-6000 mg/L, taking the molar mass differences of acetate, propionate and butyrate into account. The required build-up was achieved at UDDT-FS:OMW ratios 0:1, 1:4, 1:2 and 1:1 at both pH levels. However, the application of high OMW fractions is disadvantageous due to: 1) Logistic concerns due to collection, sorting and transportation costs of the waste from the LIHDS to the treatment site, 2) At high OMW fraction e.g. UDDT-FS:OMW ratios 1:4, 1:2 and 1:1, pH declines to very low levels, which may be toxic to *E. coli* as well all other microbial population. For practical purposes the build-up of ND-VFA in the range of 1.2-1.8 meq/g total TS added seems to be sufficient, which agrees with a ND-VFA concentration of approximately 2800-4300 mg/L, causing between 3-5 *E. coli* log inactivation in four days. This was achieved by an UDDT-FS:OMW ratio 2:1, pH$_{un}$ and a ratio of 4:1, at both pH levels.

This treatment can be beneficial in various situations including emergency cases where waste is anaerobically hydrolysed for sanitisation before disposal.

3.4 Conclusions

Experiments were conducted to investigate the potentials for UDDT-FS waste sanitisation by VFA build-up. Results showed a correlation between OMW fraction in the substrate and *E. coli* inactivation, with an increasing trend in *E. coli* inactivation being observed at increasing OMW fraction.

In co-digestion, maximum ND-VFA build-up was achieved at UDDT-FS:OMW ratio 1:1, where over 90% of the measured TVFA existed in their non-dissociated form, achieving *E. coli* inactivation to below detectable levels in 3 days. A decline in ND-VFA fraction at UDDT-FS:OMW ratios 1:2, 1:4 was attributed to an increased toxicity effect on all bacterial species, including the acidifying organisms.

ND-VFA build-up in the range of 2.0-2.5 meq/g total TS added, leading to a NF-VFA concentration of approximately 4800-6000 mg/L, was needed to achieve 10 *E. coli* log inactivation and complete *A. lumbricoides* eggs inactivation in four days. An ND-VFA build-up in the range of 1.2-1.8 meq/g TS added, agreeing with a ND-VFA concentration of approximately 2800-4300 mg/L, was needed to achieve *E. coli* log inactivation in the range of 3-5 in four days. As such, the degree of sanitisation of UDDT-FS, depends on the OMW fraction applied.

Acknowledgements

This research is funded by the Bill & Melinda Gates Foundation under the framework of SaniUp project (Stimulating local Innovation on Sanitation for the Urban Poor in Sub-Saharan Africa and South-East Asia) (OPP1029019). The authors would like to thank Ani Vabharneni, Sanergy Kenya, and DVC-ARS, Meru University, Kenya for their valuable support during this study.

References

Abdul, P., Lloyd, D. 1985. Pathogen survival during anaerobic digestion: Fatty acids inhibit anaerobic growth of Escherichia coli. *Biotechnology Lett.* , 7, 125-128.

AGROIWATECH. DeliverableD2. Analytical methods for waste water characterization and evaluation of reactor performance during anaerobic treatment. *Lettinga Associates Foundation Postbus500 6700 AM Wageningen. http://www.leaf-wageningen.nl/en/leaf.htm.*

APHA. 1995. Standard Methods for the Examination of Water and Wastewater, nineteenth ed. American Public Health Association, Washington DC, USA.

Austin, A. 2001. Health Aspects of Ecological Sanitation. Paper presented at the First International Conference on Ecological Sanitation 5-8 November 2001, Nanning, China.

Austin, L.M., Cloete, T.E. 2008. Safety aspects of handling and using fecal material from urine-diversion toilets - a field investigation. *Water Environ. Res.*, 80(4), 308-315.

Avery, L.M., Anchang, K.Y., Tumwesige, V., Strachan, N., Goude, P.J. 2014. Potential for Pathogen reduction in anaerobic digestion and biogas generation in Sub-Saharan Africa. *Biomass and Bioenergy*, 70(0), 112-124.

Battimelli, A., Carrère, H., Delgenès, J.P. 2009. Saponification of fatty slaughterhouse wastes for enhancing anaerobic biodegradability. *Bioresour. Technol.* , 100(15), 3695-3700.

Buckley, C.A., Foxon, K.M., Hawksworth, D.J., Archer, C., Pillay, S., Appleton, C., Smith, M., Rodda, N. 2008. Water Research Commission Report No. TT 356/08. Research into UD/VIDP (urine diversion ventilated improved double pit) toilets: prevalence and die-off of Ascaris ova in urine diversion waste. Durban, South Africa: Pollution Research Group, University of KwaZulu-Natal.

Byamukama, D., Kansiime, F., Mach, R.L., Farnleitner, A. 2000. Determination of Escherichia coli Contamination with Chromocult Coliform Agar Showed a High Level of Discrimination Efficiency for Differing Fecal Pollution Levels in Tropical Waters of Kampala, Uganda. *Applied and Environmentalmicrobiology*, 66(2), 864-868.

Chaggu, E.J. 2004. Sustainable Environmental Protection Using Modified Pit-Latrines. Ph.D Thesis, Wageningen University, The Netherlands.

Chen, Y., Jiang, S., Yuan, H., Zhou, Q., Gu, G. 2007. Hydrolysis and acidification of waste activated sludge at different pHs. *Water Research*, 41(3), 683-689.

Dudley, D.J., Guentzel, M.N., Ibarra, M.J., Moore, B.E., Sagik, B.P. 1980. Enumeration of potentially pathogenic bacteria from sewage sludge. *Applied Environmental Microbiology*, 39, 118-126.

Fagbohungbe, M.O., Herbert, B.M.J., Li, H., Ricketts, L., Semple, K.T. 2015. The effect of substrate to inoculum ratios on the anaerobic digestion of human faecal material. *Environmental Technology & Innovation*, 3(0), 121-129.

Farrah, S.R., Bitton, G. 1983. Bacterial survival and association with sludge flocs during aerobic and anaerobic digestion of wastewater sludge under laboratory conditions. *Applied Environmental Microbiolobilogy*, 45(1), 174-181.

Finney, M., Smullen, J., Foster, H.A., Brokx, S., Storey, D.M. 2003. Evaluation of Chromocult coliform agar for the detection and enumeration of Enterobacteriaceae from faecal samples from healthy subjects. *Journal of Microbiological Methods*, 54(3), 353-358.

Foliguet, J.M., Doncoeur, F. 1972. Inactivation in fresh and digested wastewater sludges by pasteurization. *Water Research*, 6, 1399-1407.

Fonoll, X., Astals, S., Dosta, J., Mata-Alvarez, J. 2015. Anaerobic co-digestion of sewage sludge and fruit wastes: Evaluation of the transitory states when the co-substrate is changed. *Chemical Engineering Journal*, 262(0), 1268-1274.

Frampton, E.W., Restaino, L., Blaszko, N. 1988. Evaluation of the b-glucuronidase substrate 5-bromo-4-chloro-3-indolyl-b-D-glucuronide (XGLUC) in a 24 h direct plating method for Escherichia coli. *J. Food Prot.*, 51, 402-404.

Ghimire, A., Valentino, S., Frunzo, L., Trably, E., Escudié, R., Pirozzi, F., Lens, P.N.L., Esposito, G. 2015. Biohydrogen production from food waste by coupling semi-continuous dark-photofermentation and residue post-treatment to anaerobic digestion: A synergy for energy recovery. *International Journal of Hydrogen Energy*, 40(46), 16045-16055.

Ghosh, S., Henry, M.P., Christopher, R.W. 1985. Hemicellulose conversion by anaerobic digestion. *Biomass*, 6(4), 257-269.

Gibbs, R.A., Hu, C.J., Ho, G.E., Phillips, P.A., Unkovich, l. 1995. Pathogen die-off in stored wastewater sludge. *Water Science and Technology*, 31(5–6), 91-95.

Goepfert, J., M.,, Hicks, R. 1969. Effect of Volatile Fatty Acids on Salmonella typhimurium1. *Journal of Bacteriology* 97(2), 956-958.

Jankowska, E., Chwiałkowska, J., Stodolny, M., Oleskowicz-Popiel, P. 2015. Effect of pH and retention time on volatile fatty acids production during mixed culture fermentation. *Bioresource Technology*, 190(0), 274-280.

Jensen, P.D., Ge, H., Batstone, D.J. 2011. Assessing the role of biochemical methane potential tests in determining anaerobic degradability rate and extent. *Water Sci Technol*, 64(4), 880-6.

Jiang, J., Zhang, Y., Li, K., Wang, Q., Gong, C., Li, M. 2013. Volatile fatty acids production from food waste: Effects of pH, temperature, and organic loading rate. *Bioresource Technology*, 143(0), 525-530.

Jimenez-Cisneros, B.E. 2007. Helminth ova control in wastewater and sludge for agricultural reuse, in Water and Health ,[Ed.W.O.K. Grabow],in encyclopedia of Life Support Systems(EOLSS), Developed under the Auspices of the UNESCO, Eolss Publishers, Oxford ,UK, [http://www.eolss.net] [Retrieved October 9, 2008].

Jobling Purser, B.J., Thai, S.M., Fritz, T., Esteves, S.R., Dinsdale, R.M., Guwy, A.J. 2014. An improved titration model reducing over estimation of total volatile fatty acids in anaerobic digestion of energy crop, animal slurry and food waste. *Water Research*, 61, 162-170.

Kearney, T.E., Larkin, M.J., Frost, J.P., Levett, P.N. 1993. Survival of pathogenic bacteria during mesophilic anaerobic digestion of animal waste. *Journal of Applied Microbiology*, 75:215e9.

Kim, D.-H., Kim, S.-H., Jung, K.-W., Kim, M.-S., Shin, H.-S. 2011. Effect of initial pH independent of operational pH on hydrogen fermentation of food waste. *Bioresource Technology*, 102(18), 8646-8652.

Kunte, D.P., Yeole, T.Y., Ranade, D.R. 2000. Inactivation of Vibrio cholerae during anaerobic digestion of human night soil. *Bioresource Technology*, 75(2), 149-151.

Lahav, O., Morgan, B.E. 2004. Titration methodologies for monitoring of anaerobic digestion in developing countries—a review. *Journal of Chemical Technology & Biotechnology*, 79(12), 1331-1341.

Larsen, T.A., Maurer, M. 2011. 4.07 - Source Separation and Decentralization. *Treatise on Water Science*, 203-229.

Leclerc, H., Brouzes, P. 1973. Sanitary aspects of sludge treatment. *Water Research*, 7(3), 355-360.

Lee, W.S., Chua, A.S.M., Yeoh, H.K., Ngoh, G.C. 2014. A review of the production and applications of waste-derived volatile fatty acids. *Chemical Engineering Journal*, 235, 83-99.

Lützhøft, H.-C.H., Boe, K., Fang, C., Angelidaki, I. 2014. Comparison of VFA titration procedures used for monitoring the biogas process. *Water Research*, 54, 262-272.

Manafi, M., Kneifel, W. 1989. A combined chromogenic fluorogenic medium for the simultaneous detection of total coliforms and *E. coli* in water. *Zentbl. Hyg. Umweltmed.*, 189, 225-234.

Mata-Alvarez, J.M. 1987. A dynamic simulation of a two-phase anaerobic digestion system for solid wastes. *Biotechnol. Bioeng*, 30, 844-851.

McKinney, R.E., Langley, H.E., Tomlinson, H.D. 1958. Survival of Salmonella typhosa during anaerobic digestion. I. Experimental methods and high rate digester studies. *Sewage Ind. Wastes*, 30, 1467-1477.

Moosbrugger R. E., Wentzel M. C., Ekama G. A., Marais .GvR. 1993. Weak acid/bases and pH control in anaerobic systems-a review. *Water SA 19:1-10*.

Nallathambi Gunaseelan, V. 1997. Anaerobic digestion of biomass for methane production: A review. *Biomass and Bioenergy*, 13(1–2), 83-114.

Niwagaba, C., Kulabako, R.N., Mugala, P., Jönsson, H. 2009a. Comparing microbial die-off in separately collected faeces with ash and sawdust additives. *Waste Management*, 29(7), 2214-2219.

Niwagaba, C., Nalubega, M., Vinnerås, B., Sundberg, C., Jönsson, H. 2009b. Bench-scale composting of source-separated human faeces for sanitation. *Waste Management*, 29(2), 585-589.

Olsen, J.E., Jørgensen, J.B., Nansen, P. 1985. On the reduction of Mycobacterium paratuberculosis in bovine slurry subjected to batch mesophilic or thermophilic anaerobic digestion. *Agricultural Wastes*, 13(4), 273-280.

Olsen, J.E., Larsen, H.E. 1987. Bacterial decimation times in anaerobic digestions of animal slurries. Institute of Hygiene and Microbiology, Royal Veterinary and Agricultural University, 13 Bülowsvej, DK-1870 Frederiksberg C, Denmark: http://www.researchgate.net/publication/223674948Bacterial_decimation_times_in_an aerobic_di gestions_of_animal_slurriess *Biological Wastes*.

Palmowski, L., Simons, L., Brooks, R. 2006. Ultrasonic treatment to improve anaerobic digestibility of dairy waste streams. *Water Sci. Technol.*, 53, 281-288.

Pramer, D., H. , Heukelekian, Ragotskie, R.A. 1950. Survival of tubercule bacilli in various sewage treatment processes. I. Development of a method for the quantitative recovery of mycobacteria from sewage. *Public Health Rep.* , 65, 851-859.

Prohászka, L. 1980. Antibacterial Effect of Volatile Fatty Acids in Enteric *E. coli*-infections of Rabbits. *Zentralblatt für Veterinärmedizin Reihe B*, 27(8), 631-639.

Prohászka, L. 1986. Antibacterial Mechanism of Volatile Fatty Acids in the Intestinal Tract of Pigs agains Escherichia coli. *Journal of Veterinary Medicine, Series B*, 33(1-10), 166-173.

Rajagopal, R., Lim, J.W., Mao, Y., Chen, C.-L., Wang, J.-Y. 2013. Anaerobic co-digestion of source segregated brown water (feces-without-urine) and food waste: For Singapore context. *Science of The Total Environment*, 443(0), 877-886.

Romero-Güiza, M.S., Astals, S., Chimenos, J.M., Martínez, M., Mata-Alvarez, J. 2014. Improving anaerobic digestion of pig manure by adding in the same reactor a stabilizing agent formulated with low-grade magnesium oxide. *Biomass and Bioenergy*, 67, 243-251.

Sahlström, L. 2003. A review of survival of pathogenic bacteria in organic waste used in biogas plants. *Bioresource Technology*, 87(2), 161-166.

Salsali, H.R., Parker, W.J., Sattar, S.A. 2006. Impact of concentration, temperature, and pH on inactivation of Salmonella spp. by volatile fatty acids in anaerobic digestion. *Canadian Journal of Microbiology*, 52: 279-286.

Sherpa, A.M., Byamukama, D., Shrestha, R.R., Haberl, R., March, R.L., Farnleitner, A.H. 2009. Use of faecal pollution indicators to estimate pathogen die-off conditions in source-separated faeces in Kathmandu Valley, Nepal. *J. Water Health*, 7(1), 97-107.

Van Lier, J.B., Grolle, K.C.F., Frijters, C.T.M.J., Stams, A.J.M., Lettinga, G. 1993. Effect of acetate, propionate and butyrate on the thermophilic anaerobic degradation of propionate in methanogenic sludge and defined cultures. *Appl. and Environ. Microbiol*, 57, 1003-1011.

Van Lier, J.B., Mahmoud, N., Zeeman, G. 2008. Biological wastewater treatment: Principles, modelling and design. Chapter 16: Anaerobic waste water treatment. IWA publishing.

Vinnerås, B. 2007. Comparison of composting, storage and urea treatment for sanitising of faecal matter and manure. *Bioresource Technology*, 98(17), 3317-3321.

Wang, K., Yin, J., Shen, D., Li, N. 2014. Anaerobic digestion of food waste for volatile fatty acids (VFAs) production with different types of inoculum: Effect of pH. *Bioresource Technology*, 161(0), 395-401.

WHO. 2006. Guidelines for the safe use of wastewater, excreta and greywater. Volume 4. Excreta and greywateruse in agriculture; ISBN 92 4 154685 9. http://whqlibdoc.who.int/publications/2006/9241546859_eng.pdf.

Zhang, B., Zhang, L.L., Zhang, S.C., Shi, H.Z., Cai, W.M. 2005. The influence of pH on hydrolysis and acidogenesis of kitchen wastes in two-phase anaerobic digestion. *Environmental technology*, 3, 329-339.

Chapter 4: Volatile fatty acids (VFA) build-up and its effect on *E. coli* inactivation during excreta digestion in single-stage and two-stage systems

This Chapter is based on a paper: Riungu J., Ronteltap, M., van Lier, J.B. 2018. Volatile fatty acids (VFA) build-up and its effect on *E. coli* inactivation during excreta digestion in single-stage and two- stage systems. *Journal of Water Sanitation and Hygiene for Development. 10.2166/washdev.2018.160.*

Volatile fatty acids (VFA) build-up and its effect on *E. coli* inactivation during excreta digestion in single-stage and two-stage systems

Abstract

Digestion and co-digestion of faecal matter collected from Urine Diverting Dehydrating Toilet Faeces (UDDT-FS) and mixed Organic Market Waste (OMW) was studied in single stage pilot scale mesophilic plug-flow anaerobic reactors at UDDT-FS:OMW ratios 4:1 and 1:0. *Escherichia coli* (*E. coli*) inactivation and Volatile Fatty Acids (VFA) build-up was monitored at sampling points located along the reactor profile. *E. coli* inactivation achieved in digestion of UDDT-FS: OMW ratio of 4:1, 12% TS was 2.3 log times higher than that achieved in UDDT-FS: OMW ratio of 1:0.

In subsequent trials, a two-stage reactor was studied, applying a UDDT-FS:OMW ratio of 4:1 and 10% or 12% TS slurry concentrations. Highest VFA concentrations of 16.3±1.3 g/l were obtained at a pH of 4.9 in the hydrolysis/acidogenesis reactor, applying a UDDT-FS:OMW ratio of 4:1 and 12% TS, corresponding to a non-dissociated (ND)-VFA concentration of 6.9±2.0 g/l. The corresponding decay rate reached a value of 1.6 /d. In the subsequent methanogenic plug-flow reactor, a decay rate of 1.1/d was attained within the first third part of the reactor length, which declined to 0.6/d within the last third part of the reactor length. Results show that a two-stage system is an efficient way to enhance pathogen inactivation during anaerobic digestion.

4.1 Introduction

Ecological sanitation concepts have been developed due to the growing need for improved onsite sanitation systems aiming at the protection of human and environmental health (Esrey, 2001; Niwagaba *et al.*, 2009a). Urine Diverting Dehydrating Toilets (UDDTs) fit well into this concept, especially in densely populated, low lying settlements (Katukiza *et al.*, 2012; Niwagaba *et al.*, 2009a; Schouten & Mathenge, 2010). The technology has been adopted by Sanergy (Nairobi, Kenya), a company working on sanitation in LIHDS. Currently, from Mukuru Kwa Njenga and Mukuru Kwa Reuben LIHDS, approximately seven tonnes UDDT-faeces (UDDT-FS) are delivered per day to the central treatment plant, located 50 km from the city centre. Key concern is digestion and sanitisation of the waste, as the addition of ash and sawdust after toilet use is insufficient for pathogen inactivation (Niwagaba *et al.*, 2009a).

Anaerobic digestion (AD) provides a cost effective and energy saving alternative for waste treatment (Avery *et al.*, 2014; Fonoll *et al.*, 2015; Nallathambi Gunaseelan, 1997; Romero-Güiza *et al.*, 2014). Anaerobic systems can be applied at any scale and almost any place, whereas the effluent is stabilised with a good fertiliser value (Pabón-Pereira *et al.*, 2014; Van Lier *et al.*, 2008). Key reported drawback, however, is insufficient pathogen inactivation with solid and liquid digestate containing high levels of pathogenic organisms such as *Salmonella, Shigella, Campylobacter jejuni, Clostridium perfringens, Enterococcus species* and V*ibrio cholera* (Chaggu, 2004; Chen *et al.*, 2012; Fagbohungbe *et al.*, 2015; Horan *et al.*, 2004; Kunte *et al.*, 2000; Massé *et al.*, 2011), As such, the poor microbial quality of the digested solids may lead to transmission of enteric diseases when applied to agricultural land (Pennington, 2001; Smith *et al.*, 2005).

During anaerobic digestion, temperature and time play a key role in pathogen inactivation (Gibbs *et al.*, 1995; Olsen *et al.*, 1985; Olsen & Larsen, 1987; Smith *et al.*, 2005), as does reactor configuration (Kearney *et al.*, 1993; Olsen *et al.*, 1985). In addition, pH and VFA concentration in the reactor broth are an indication for bacterial survival (Abdul & Lloyd, 1985; Farrah & Bitton, 1983; Sahlström, 2003). At a low reactor pH, the same amount of VFAs lead to a higher fraction of non-dissociated VFAs (ND-VFAs), which may result in higher microbial decay: ND-VFAs pass freely bacterial cell walls by passive diffusion and affect the internal pH (Jiang *et al.*, 2013; Riungu *et al.*, 2018b; Wang *et al.*, 2014a). However, during the digestion of sewage sludge the high buffer capacity limits pH changes (Fonoll *et al.*, 2015; Franke-Whittle *et al.*, 2014; Gallert *et al.*, 1998; Murto *et al.*, 2004) and hence reduces the options of using ND-VFAs for pathogen inactivation. By co-digesting human waste (UDDT-FS) with mixed organic market waste (OMW), acid formation is enhanced, since OMW is carbohydrate rich and easily hydrolysable (Gómez *et al.*, 2006; Lim *et al.*, 2008).

Enhanced build-up of total VFA (TVFA) concentrations during co-digestion of sewage sludge and other organic waste can be achieved by inhibition of methanogenesis (Wang *et al.*, 2014), through use of a two-stage reactor system, where hydrolysis/acidogenesis and methanogenesis are separated. The different species of micro-organisms involved in the AD process can be divided into two main groups of bacteria, namely organic acid producing and organic acid consuming or methane forming microorganisms (Rincón *et al.*, 2008). They operate under different pH conditions: whereas the optimal pH for acidogenic bacteria activity ranges between 5 and 7 (Fang & Liu, 2002; Guo *et al.*, 2010; Liu *et al.*, 2006; Noike *et al.*, 2005), methanogenic activity requires a minimum pH of 6.5 (Wang *et al.*, 2014b; Yuan *et al.*, 2006). Key drawback in the two-stage reactor is the high VFA concentration in the acidogenic reactor, which requires pH correction for stable methanogenesis (Zuo *et al.*, 2014). Yet, the low pH and high VFA concentrations create very good pathogen inactivating conditions. Hence, an optimum must be found between good pathogen removal and well-functioning methanogenic digestion. In practice, the latter can be achieved by recycling part of the digestate upfront to be mixed with the acidified UDDT-FS-OMW.

In Chapter 3, we evaluated the effect of UDDT-FS and OMW mix ratios on VFA build-up and *E. coli* inactivation in laboratory scale batch anaerobic reactors, within a retention time of four days. *E. coli* inactivation was a function of the OMW fraction in the substrate, increasing as the fraction increased (Riungu *et al.*, 2018a). The ratio appropriateness depends on the required degree of sanitisation, final pH values in the final digestate, and obviously, the availability of OMW.

This study evaluated the potential for pathogen inactivation in anaerobic digestion, co-digesting UDDT-FS and OMW, using pilot scale plug-flow reactors. In particular, study results give a comparison of *E. coli* inactivation from single and two-stage plug-flow reactors.

4.2   Materials and Methods

4.2.1 UDDT-Faeces waste samples

UDDT-FS samples used for this study were obtained from the Fresh Life© urine diverting dry (UDDT) toilets within Mukuru Kwa Njenga/ Mukuru Kwa Reuben LIHDS, Kenya. The Fresh Life© toilets are fabricated and installed by a social enterprise, Sanergy, in collaboration with entrepreneurs in the informal settlements who maintain them. The toilets are provided on a pay-and-use basis, charging approximately 0.05 euro/use and have an average user load of 50 persons/day. Within each toilet facility, a 30 L container is used for waste collection, with approximately 10 g sawdust added after every toilet use. The toilets are emptied on daily basis, where used containers are replaced with clean ones.

From a batch consisting of about 60 containers, ten containers were randomly selected after which mixing of the contents was done in order to obtain a homogeneous mix. 15 kg UDDT-FS was then drawn and further mixing was done in order to homogenise the sample.

4.2.2 Mixed Organic Market Waste samples (OMW)

Mixed organic market waste (OMW) was obtained from Mukuru Kwa Njenga and Mukuru Kwa Reuben LIHDS. About 20 kg of the waste was collected daily and contained food waste, vegetable waste and fruit waste, in about equal proportions. Size reduction of OMW substrates for pilot scale tests was achieved by manual chopping to about 1 cm size. Table 4.1 shows the characteristics of the material used in this study.

Table 4.1: Characterisation of UDDT-FS and OMW used in the study (Adopted from Riungu et. al., 2018)

	UDDT-FS		OMW	
	Value	STDEV	Value	STDEV
TS (% wgt)	24.5	3.8	17.9	1.6
Moisture content	75.5	3.8	80.7	4.1
VS (% wgt)	20.1	3.5	16.9	4.4
E. coli (CFU/g TS)	1.7E+09	5.3E+08	2.7E+05	7.4E+04
Ascaris eggs	Not detected		Not detected	

4.2.3 Pilot scale AD experiments

Two sets of reactors were used, namely a single stage reactor (R_s) and a two-stage reactor (R_{am}) comprising of a hydrolysis/ acidogenic reactor (R_a) and a methanogenic reactor (R_m).

Experiments were conducted at UDDT-FS:OMW ratio 4:1, at 10% and 12% total solids (TS) concentrations. Substrate concentration selection was based on a series of laboratory scale batch-tests derived experimental data on the effect of substrate concentration on pathogen inactivation (Riungu *et al.*, 2018a). Research aimed at treating the highest possible TS concentration that can freely flow through the plug-flow reactor without the necessity of using pumps.

Hydrolysis reactor design

The single stage reactors R_a's were fabricated from 30 L plastic containers, with a working volume of 20 L. These reactors were equipped with a cover, incorporated with two separate ports, i.e. a feeding port and a port fixed with a manual stirring mechanism, whereas the bottom of each reactor was equipped with a discharge/ effluent valve.

Plug flow reactor design

Six plug flow digesters (Figure 4.1) were constructed using 175 L tubular polyethylene bags. Each of the bags had a diameter of 30 cm and a length of 2.1 m and the polyethylene material had a thickness of 0.2 mm. Produced biogas flowed by pressure to a 175 L biogas storage bag that was installed directly above each reactor. In addition, three separate ports were incorporated onto each bag namely: inlet port (SP_1); a sampling port (SP_2) at 0.7 m digester length; a gas discharge port at 1.4 m digester length; and an effluent/discharge port (SP_3) at 2.1 m digester length. A total solids retention time (SRT) of 29 days was maintained for the anaerobic digestion process.

Figure 4.1: Plug flow digester layout: reactors on the floor, biogas collection bags directly above; sampling points are indicated (SP1, 2 and 3)

4.2.3.1 Plug flow reactor start-up and operation in single substrate and co-digestion experiments

For smooth startup, reactors were inoculated using inoculum obtained from fixed dome anaerobic digesters (operated by Umande Trust, Nairobi, Kenya, https://umande.org/). The inoculum upon collection was incubated for one week to methanise any organic matter before use.

The six plug flow reactors D_1, D_2, D_3, D_4, D_5 and D_6, were divided into two groups (D_1, D_3 and D_5, and D_2, D_4 and D_6), representing two treatment groups in single substrate digestion of UDDT-FS:OMW ratio 1:0 at 12% TS and 10% TS respectively. About 5 L/day of the appropriate substrate was fed each to respective digesters every morning. Digestion of the digesters was achieved after 1.5 months, and sample collection and analysis commenced and continued for a further 9 weeks.

Co-digestion experiments with UDDT-FS:OMW ratio 4:1 at 12% TS concentration commenced 15 weeks after the start-up. The experiments aimed for comparing pathogen inactivation in single (R_s) and two-stage (R_{am}) anaerobic digestion processes. Three replications of two treatments groups R_{am} and R_s were set up, with D_1, D_3 and D_5 being R_{am}'s and D_2, D_4 and D_6 being R_s's. Each morning, UDDT-FS:OMW ratio of 4:1, 12% TS concentration was prepared after which 5 L of the substrate was fed into the R_s reactors. In R_{am} reactor setup, effluent from R_a acted as influent to the R_m. Details on design of R_a are provided on section 2.3.1 Two R_a's were operated in parallel and every morning, 5 L of effluent was drawn from each and mixed. pH of the mixture was adjusted to the range of 5.8-6.2 using effluent from R_m reactors. Thereafter, 6 L of the mix was fed to each of the three R_m's (D_1, D_3 and D_5) every morning.

Finally, the concentration of the feed into R_{am} was reduced to 10% TS. Thereafter, 100 mL of R_s, R_a and R_m's influent and effluent were sampled for analysis of moisture content, total solids, volatile solids, *E. coli* and VFA.

4.2.4 Analytical procedures

4.2.4.1 Total solids and volatile solids

Total solids and volatile solids analysis were conducted according to the gravimetric method (SM-2540D and SM-2540E), as outlined in the Standard Methods for the Examination of Water and Wastewater (APHA, 1995b). pH measurement was done using a calibrated analogue pH/ORP meter (model HI8314-S/N 08586318).

4.2.4.2 VFA measurements

The method used is based on esterification of the carboxylic acids present in the sample and subsequent determination of the esters by the ferric hydroxamate reaction (DR 2800 Hach, June 2007 Edition). The method has a measuring range of 27-2800 mL/L. As such, homogenised samples were serially diluted (10^{-1} to 10^{-6}) with de-ionised water to get the correct measuring range.

From the TVFA concentration, the fraction of ND-VFAs was calculated. VFAs are commonly considered to constitute a single weak-acid system with a single equilibrium constant Ka because of the similarity of their pK values (Lahav & Morgan, 2004; Moosbrugger R. E. *et al.*, 1993). Therefore:

$$((H^+).(A^-))/(HA) = Ka\dots\dots\dots\dots \text{(Equation 1)}$$

$$pH = pKa + ^{10}log\left(\frac{A^-}{HA}\right)\dots \text{(Equation 2)}$$

$$A_T = (HA) + (A^-)\dots\dots\dots\dots \text{(Equation 3)}$$

Where: A_T = total VFA species concentration (mg/L), HA represents the acidic, protonated species and A⁻ the ionised form of each acid.

Similarly, TOC and COD measurements were done using protocols adopted from Hach spectrophotometer, DR 2800.

4.2.4.3 *E. coli* enumeration

E. coli, one of the indicator organism for possible use of digestate coming from faecal matter, in agriculture was used as an indicator organism for pathogen inactivation. It's enumeration was done using surface plate technique with Chromocult Coliform Agar (Chromocult: Merck, Darmstadt, Germany) (Byamukama *et al.*, 2000; Mawioo *et al.*, 2016). The first order reaction coefficients for *E. coli* inactivation were calculated using the Chick-Watson model that expresses the rate of inactivation of micro-organisms by a first order chemical reaction (Gerba, 2008).

$$\ln\left(\frac{C_t}{C_0}\right) = -kt \dots\dots\dots (Equation\ 4)$$

Where:
C_t = *Number of micro − organisms at time t*
C_0 = *Number of micro − organisms at time 0*
k = *decay rate*
t = *time*

Using results, *E. coli* inactivation (−ln(ct/co)) was plotted against time (Figure 4.6).

4.3 Results and discussion

4.3.1 Evaluation of the performance of single stage reactor (R_s) system

ND-VFA profiles

An evaluation of the performance of single stage plug flow reactor (R_s) was carried out using UDDT-FS:OMW ratio 4:1, 12% TS ($R_{s-4:1,\ 12\%}$), UDDT-FS:OMW ratio 1:0, 12% TS ($R_{s-1:0,\ 12\%}$) and UDDT-FS:OMW ratio 1:0, 10% TS ($R_{s-1:0,\ 10\%}$) systems, with results shown on Figure 4.2 (a-c). Among the tested substrates, co-digestion ($R_{s-4:1,\ 12\%}$) showed highest TVFA and ND-VFA build-up, with a 4-fold increase in ND-VFA and 3.2-fold increase in TVFA build-up being observed between influent (SP₁) and SP₂ sampling points. However, in $R_{s-1:0,\ 12\%}$ and

$R_{s-1:0, 10\%}$, a 6 and 6.5-fold decline in ND-VFA concentration was observed between sampling points SP_1 and SP_2 respectively, owing to an increase in the local pH. OMW, associated with rapid hydrolysis (Riungu *et al.*, 2018a; Zhang *et al.*, 2008) , enhanced the VFA build up in the digestion medium when used as co-substrate, and thus increased the ND-VFA concentration, particularly when a concomitant pH drop is observed (Riungu *et al.*, 2018a). However, a sharp decline in TVFA and ND-VFA concentration was observed between SP_2 and SP_3, which indicated proper methanogenic conditions in the final stages of the plug-flow reactor reaching pH values of 7.5. Decline in ND-VFA concentration in $R_{s-1:0, 12\%}$ and $R_{s-1:0, 10\%}$ reactors along the reactor length may be attributed to the high buffer capacity of UDDT-FS substrate and prevailing methanogenic conditions. High buffer capacity of the UDDT-FS substrate might be attributed to the occasional wrong toilet use, collecting both urine and faeces in the same vessel, resulting in increased ammonium bicarbonate concentrations.

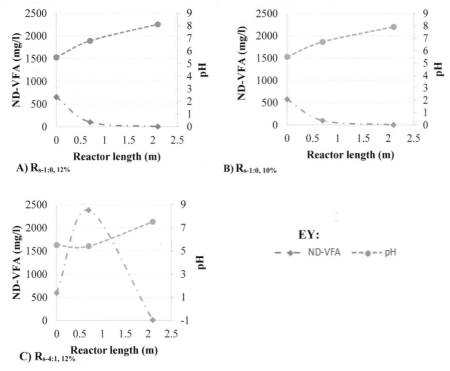

Figure 4.2: Development of ND-VFA (blue) and pH (red) along the reactor length

The effluent pH in single-stage single substrate and co-digestion experiments reactor set-ups was comparable and within the optimal range for methanogenic bacteria, i.e. 7.5-8.1 (Figure 4.2a-c). pH control along these reactor profiles was self-regulatory. Between SP_2 and SP_3, whereas in single substrate reactors a gradual increase in pH was observed, the co-digestion reactor ($R_{s-4:1,\ 12\%}$) showed similar pH in the influent (SP_1), and the first sampling point (SP_2) followed by a sharp increase between SP_2 and SP_3 (see Figure 4.2c). The low pH at SP_2 resulted from OMW hydrolysis/acidification, which emphasises the importance of a proper UDDT-FS:OMW ratio, avoiding full system acidification and potential failure. In full-scale systems, recycle flows may be used for pH regulations preserving methanogenic conditions in the final stage. Use of the recycle stream for pH adjustment in the two-stage reactor system was sufficient to guarantee methanogenic conditions in the plug-flow reactors.

E. coli log inactivation in single substrate digestion

Figure 4.3 depicts *E. coli* log inactivation trends in $R_{s-4:1,\ 12\%}$, $R_{s-1:0,12\%}$ and $R_{s-1:0,10\%}$ at the three sampling points SP_1, SP_2 and SP_3, located at 0, 0.7 and 2.1 m of the reactor length, respectively. The higher pathogen inactivation shown in $R_{s-4:1,\ 12\%}$ (Figure 4.3C) coincides with the prevailing higher maximum ND-VFA concentrations as a consequence of increased OMW hydrolysis/acidification. As discussed in section 3.1.1, between sampling points SP_1 to SP_2, ND-VFA declined in $R_{s-1:0,\ 12\%}$ and $R_{s-1:0,\ 10\%}$, with an increase being observed in $R_{s-4:1,\ 12\%}$ system. The increase in ND-VFA allowed more contact time of the pathogens to the high ND-VFA concentrations, consequently leading to higher inactivation.

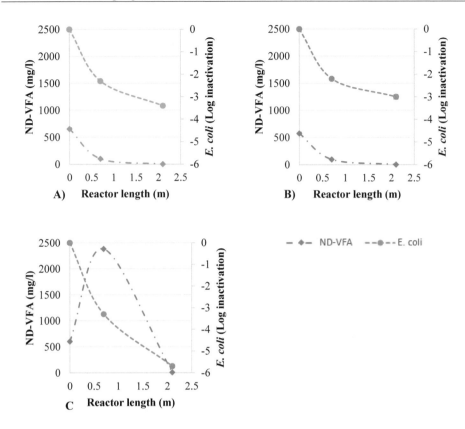

Figure 4.3: *E. coli* **log inactivation (red) and the production of non-dissociated VFAs (blue) in single stage anaerobic digestion of UDDT-FS at 10% and co-digestion of UDDT-FS and OMW at 12% TS**

The *E. coli* removal in the two stage co-digestion reactors, applying a UDDT-FS:OMW ratio of 4:1 and 12% TS (Section 3.2.2), showed an 8.0 log inactivation, whereas only a 5.7 log inactivation was achieved in the single stage co-digestion reactor at 12% TS. Results indicate that the two-stage reactor is about 200 times more effective in removing the *E. coli* indicator organism.

4.3.2 Co-digestion of UDDT-FS and OMW in a two-stage reactor (R_{am}) system

Co-digestion of UDDT-FS and OMW ratio 4:1 (UDDT-FS:OMW ratio 4:1) was evaluated in a two-stage reactor system, 10% TS ($R_{m-4:1,\ 10\%}$) and 12% TS ($R_{m-4:1,\ 12\%}$)

concentration whereas single stage reactor digesting UDDT-FS:OMW ratio 4:1 ($R_{s-4:1,\ 10\%}$) acted as control reactor. Details on design and operation of the reactors have been discussed in section 2.2.

Volatile fatty acids and pH changes

Table 4.2 shows the trend in Total Volatile Fatty Acids (TVFA), Non-Dissociated Fatty acids (ND-VFA) and pH in the single stage and two-stage co-digestion reactor systems.

Table 4.2: Variation in TVFA, ND-VFA and pH in R_m and R_s reactors.

	Reactor	Parameter	SP$_1$	SP$_2$	SP$_3$
Co-digestion UDDT-FS:OMW ratio 4:1	$R_{m-4:1,\ 12\%}$	TVFA (mg/l)	15685±1772	10526±844	1575±607
		ND-VFA (mg/l)	800±112	286±68	1.7±0.2
		ND-VFA (%)	5.1±0.6	2.7±0.6	0.1
		pH	6.4±0.1	6.4±0.1	7.8±0.1
	$R_{m-4:1,\ 10\%}$	TVFA (mg/l)	12347±887	8702±72	1744±101
		ND-VFA (mg/l)	660±311	281±49	1.6±0.3
		ND-VFA (%)	3.5±2	3.2±0.6	0.1
		pH	6.3±0.1	6.2±0.1	7.8±0.1
	$R_{s-4:1,12\%}$	TVFA (mg/l)	3844±679	12121±1153	2629±326
		ND-VFA (Mg/l)	599.4±150	2379±409	5±1.2
		ND-VFA (%)	15.8±3.4	19.6±2.8	0.2
		pH	5.4±0.1	5.4±0.1	7.5±0.1
		Temperature (°C)		30.1±0.3	

$R_{m-4:1,\ 12\%}$ and $R_{m-4:1,\ 10\%}$ showed similar trends in TVFA, ND-VFA and pH during the entire experimental period. The influent to the methanogenic reactors showed high TVFA concentrations, attributed to biomass pre-hydrolysis/acidification in the acidogenic reactors (Table 4.2). However, ND-VFA concentration in two-stage reactors was low (e.g. 0.8 g/l for $R_{m-4:1,\ 12\%}$) compared to acidogenic reactor (R_a) effluent (6.9±2.0 g/L), due to the buffer effect of the recycle stream used for pH adjustment. A declining trend was observed in both TVFA and ND-VFA along the reactor length. For example, at the mid sampling point (SP$_2$) TVFA and ND-VFAs concentrations in $R_{m-4:1,\ 12\%}$ were 10.5 and 0.3 g/l respectively, distinctly lower than its corresponding influent (SP$_1$) concentration of 15.6 g/l and 0.8 g/l, respectively. With an average pH of 6.4 at the first 2 sampling points, the distinct drop in TVFA can be attributed to prevailing methanogenesis (Goepfert & Hicks, 1969a). Whereas in the two-stage system hydrolysis/acidogenesis was clearly located in the separate R_a reactor, in the single stage plug flow reactor, hydrolysis/acidogenesis prevailed between sampling points SP_1 and SP_2.

E. coli inactivation along reactor profile in $R_{am-4:1,\ 10\%}$ and $R_{am-4:1,\ 12\%}$

$R_{am-4:1,\ 12\%}$ depicted 8.0 log *E. coli* inactivation, slightly higher than the corresponding value of 7.3 log attained in the $R_{am-4:1,\ 10\%}$ system (Figure 4.4, A & B).

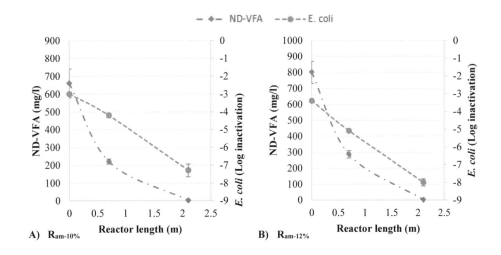

Figure 4.4: ND-VFA concentrations and *E. coli* inactivation variation along the reactor profile: A) $R_{am-10\%}$ and B) $R_{am-12\%}$

The observed improved inactivation can likely be attributed to the initial hydrolytic/ acidogenic phase of the R_a reactor that depicted an average of 3.4 and 3.0 *E. coli* log inactivation in $R_{am-4:1,\ 12\%}$ and $R_{am-4:1,\ 10\%}$, respectively, corresponding to decay rates of 1.6 and 1.7/day respectively (using Eq. 4). *E. coli* log inactivation in $R_{am-4:1,\ 12\%}$ and $R_{am-4:1,\ 10\%}$ systems depicted a similar trend along the reactor length.

E. coli inactivation progressed along the reactor length, with highest inactivation being achieved within the first one third of methanogenic reactor length. Between SP_1 and SP_2, a 4.2 and 5.1 *E. coli* log inactivation was achieved in $R_{am-4:1,\ 10\%}$ and in $R_{am-4:1,\ 12\%}$ respectively, corresponding to decay rates of 1.1 and 0.9 /day respectively. Overall, in the entire $R_{am-4:1,\ 12\%}$ and $R_{am-4:1,\ 10\%}$ system, an 8.0 and 7.2 *E. coli* log inactivation was achieved at SP_3 (effluent), corresponding to a decay rate of 0.6 in both cases. Moreover, ND-VFA calculated concentration in $R_{am-4:1,\ 12\%}$ and $R_{am-4:1,\ 10\%}$ systems showed a declining trend along the reactor length (see

Figure 4.4). Apparently, the decay rate 'k' is highest under high ND-VFA conditions and levels off when ND-VFA drops and/or pH increases. Under methanogenic conditions the k-value may be governed by microbial 'predation' and chemical interactions, reported to play role in pathogen inactivation (Smith *et al.*, 2005).

The *E. coli* decay rates obtained in this study are comparable to related studies as shown on Table 4.3.

Table 4.3: Comparing *E. coli* removal with related literature results

Substrate	Temp (°C)	SRT (Day)	Log removal	k_d (/d)	Digestion method	Reference
UDDT-FS &OMW (12% TS)	30	4	3.4	0.9	CSTR fed	This study
UDDT-FS &OMW (10% TS)	30	4	3.0	0.8	CSTR fed	This study
UDDT-FS (12% TS)	30	29	3.4	0.1	Single stage plug flow	This study
UDDT-FS (10% TS)	30	29	3	0.1	Single stage plug flow	This study
UDDT-FS&OMW (12% TS)	30	29	8	0.3	Two-stage plug flow	This study
UDDT-FS&OMW (10% TS)	30	29	7.2	0.3	Two-stage plug flow	This study
UDDT-FS &OMW	30	29	5.7	0.2	Single stage plug flow	This study
Cattle slurry	35	15	6.5	0.4	Batch	(Steffen *et al.*, 1998)
Liquid sewage sludge	35	15-20	0.5-2.0	0.0-0.1	Batch	(Smith *et al.*, 2005)
Sewage Sludge	35	21	1.5-1.7	0.1	Semi-continuous	(Horan *et al.*, 2004)
Beef cattle slurry	35	7	6.5	0.9	Batch	(Kearney *et al.*, 1993a)
Beef cattle slurry	35	8	4.25	0.5	Semi-continuous	(Kearney *et al.*, 1993a)
Sewage sludge	37	21	1-2	0.1	Batch	(Higgins *et al.*, 2007)
Sewage sludge	37	21	2.0	0.1	Batch	(Higgins *et al.*, 2007)
Swine manure	24	7	2.8-2.9	0.4	SBR	(Massé *et al.*, 2011)
Cattle slurry	18e25	25	6.5	0.3	Batch	(Santha *et al.*, 2006)
Swine slurry	20	20	2-5	0.1-0.2	SBR	(Côté *et al.*, 2006)

The observed high Kd values observed in our present study may be attributed to two factors: 1) optimal activity of hydrolytic/acidogenic bacteria and concomitantly suppressing alkalinity regeneration by methanogenesis in the R_a stage of the R_{am} system, resulting in high levels of ND-VFA. 2) OMW contains appreciable amounts of fats that are easily hydrolysable to long chain fatty acids, which may impose additional toxic effects on micro-organisms involved in the AD process (Angeriz-Campoy *et al.*, 2015; Silva *et al.*, 2014).

4.4 Conclusions

This study evaluated the technical feasibility of pathogen inactivation during digestion and co-digestion of UDDT-FS and UDDT-FS-OMW mixtures. All waste substrates were obtained from Mukuru Kwa Njenga and Mukuru Kwa Reuben LIHDS, Nairobi, Kenya. Results showed that co-digesting UDDT-FS and OMW at a ratio of 4:1 in a two-stage reactor enhances sanitisation as shown by assessing *E. coli* levels along the reactor length. *E. coli* inactivation of 8.0 log units was achieved within 29 days SRT. Rapid ND-VFA build-up was achieved from the mixed waste substrate, especially within the separate completely mixed hydrolysis reactor, where ND-VFA build-up between 5200-6500 mg/L achieved 3.4 *E. coli* log inactivation in four days. An up to 5.1 log inactivation was achieved within the first one third of reactor length of the plug-flow reactor, agreeing with an SRT of 11 days.

Acknowledgements

This research is funded by the Bill & Melinda Gates Foundation under the framework of SaniUp project (Stimulating local Innovation on Sanitation for the Urban Poor in Sub-Saharan Africa and South-East Asia) (OPP1029019). The authors would like to thank Ani Vabharneni, Sanergy Kenya, and DVC-ARS, Meru University, Kenya for their valuable support during this study.

References

Abdul, P., Lloyd, D. 1985. Pathogen survival during anaerobic digestion: Fatty acids inhibit anaerobic growth of Escherichia coli. *Biotechnology Lett.* , 7, 125-128.

Angeriz-Campoy, R., Álvarez-Gallego, C.J., Romero-García, L.I. 2015. Thermophilic anaerobic co-digestion of organic fraction of municipal solid waste (OFMSW) with food waste (FW): Enhancement of bio-hydrogen production. *Bioresource Technology*, 194, 291-296.

APHA. 1995. *Standard methods, 19 edn American Public Health Association*, Washington, DC. .

Avery, L.M., Anchang, K.Y., Tumwesige, V., Strachan, N., Goude, P.J. 2014. Potential for Pathogen reduction in anaerobic digestion and biogas generation in Sub-Saharan Africa. *Biomass and Bioenergy*, 70(0), 112-124.

Byamukama, D., Kansiime, F., Mach, R.L., Farnleitner, A. 2000. Determination of Escherichia coli Contamination with Chromocult Coliform Agar Showed a High Level of Discrimination Efficiency for Differing Fecal Pollution Levels in Tropical Waters of Kampala, Uganda. *Applied and Environmentalmicrobiology*, 66(2), 864-868.

Chaggu, E.J. 2004. Sustainable Environmental Protection Using Modified Pit-Latrines. Ph.D Thesis, Wageningen University, The Netherlands.

Chen, Y., Fu, B., Wang, Y., Jiang, Q., Liu, H. 2012. Reactor performance and bacterial pathogen removal in response to sludge retention time in a mesophilic anaerobic digester treating sewage sludge. *Bioresource Technology*, 106(Supplement C), 20-26.

Côté, C., Massé, D.I., Quessy, S. 2006. Reduction of indicator and pathogenic microorganisms by psychrophilic anaerobic digestion in swine slurries. *Bioresource Technology*, 97(4), 686-691.

DR 2800 Hach. June 2007 Edition. Spectrophotometer procedures manual 2 Catalog Number DOC022.53.00725.

Esrey, S.A. 2001. Towards the recycling society: ecological sanitation - closing the loop to food security. *Wat Sci. Technol*, 43 (4), 177-187.

Fagbohungbe, M.O., Herbert, B.M.J., Li, H., Ricketts, L., Semple, K.T. 2015. The effect of substrate to inoculum ratios on the anaerobic digestion of human faecal material. *Environmental Technology & Innovation*, 3(0), 121-129.

Fang, H.H.P., Liu, H. 2002. Effect of pH on hydrogen production from glucose by a mixed culture. *Bioresource Technology*, 82(1), 87-93.

Farrah, S.R., Bitton, G. 1983. Bacterial survival and association with sludge flocs during aerobic and anaerobic digestion of wastewater sludge under laboratory conditions. *Applied Environmental Microbiolobilogy*, 45(1), 174-181.

Fonoll, X., Astals, S., Dosta, J., Mata-Alvarez, J. 2015. Anaerobic co-digestion of sewage sludge and fruit wastes: Evaluation of the transitory states when the co-substrate is changed. *Chemical Engineering Journal*, 262(0), 1268-1274.

Franke-Whittle, I.H., Walter, A., Ebner, C., Insam, H. 2014. Investigation into the effect of high concentrations of volatile fatty acids in anaerobic digestion on methanogenic communities. *Waste Management*, 34(11), 2080-2089.

Gallert, C., Bauer, S., Winter, J. 1998. Effect of ammonia on the anaerobic degradation of protein by a mesophilic and thermophilic biowaste population. . *Applied Microbiology and Biotechnology* (50), 495-501.

Gerba, C.P. 2008. Biological wastewater treatment: Principles, modelling and design. Chapter 16: Anaerobic waste water treatment. IWA publishing.

Gibbs, R.A., Hu, C.J., Ho, G.E., Phillips, P.A., Unkovich, l. 1995. Pathogen die-off in stored wastewater sludge. *Water Science and Technology*, 31(5–6), 91-95.

Goepfert, J., M., , Hicks, R. 1969. Effect of Volatile Fatty Acids on Salmonella typhimurium. *Journal of biotechnology 97(2), 956-958.*

Gómez, X., Morán, A., Cuetos, M.J., Sánchez, M.E. 2006. The production of hydrogen by dark fermentation of municipal solid wastes and slaughterhouse waste: A two-phase process. *Journal of Power Sources*, 157(2), 727-732.

Guo, X.M., Trably, E., Latrille, E., Carrère, H., Steyer, J.-P. 2010. Hydrogen production from agricultural waste by dark fermentation: A review. *International Journal of Hydrogen Energy*, 35(19), 10660-10673.

Higgins, M.J., Chen, Y.-C., Murthy, S.N., Hendrickson, D., Farrel, J., Schafer, P. 2007. Reactivation and growth of non-culturable indicator bacteria in anaerobically digested biosolids after centrifuge dewatering. *Water Research*, 41(3), 665-673.

Horan, N.J., Fletcher, L., Betmal, S.M., Wilks, S.A., Keevil, C.W. 2004. Die-off of enteric bacterial pathogens during mesophilic anaerobic digestion. *Water Research*, 38(5), 1113-1120.

Jiang, J., Zhang, Y., Li, K., Wang, Q., Gong, C., Li, M. 2013. Volatile fatty acids production from food waste: Effects of pH, temperature, and organic loading rate. *Bioresource Technology*, 143(0), 525-530.

Katukiza, A.Y., Ronteltap, M., Niwagaba, C.B., Foppen, J.W.A., Kansiime, F., Lens, P.N.L. 2012. Sustainable sanitation technology options for urban slums. *Biotechnology Advances*, 30(5), 964-978.

Kearney, T.E., Larkin, M.J., Frost, J.P., Levett, P.N. 1993. Survival of pathogenic bacteria during mesophilic anaerobic digestion of animal waste. *Journal of Applied Microbiology*, 75:215e9.

Kearney, T.E., Larkin, M.J., Levett, P.N. 1993a. The effect of slurry storage and anaerobic digestion on survival of pathogenic bacteria. *Applied Microbiology and Biotechnology*, 74, 86-93.

Kunte, D.P., Yeole, T.Y., Ranade, D.R. 2000. Inactivation of Vibrio cholerae during anaerobic digestion of human night soil. *Bioresource Technology*, 75(2), 149-151.

Lahav, O., Morgan, B.E. 2004. Titration methodologies for monitoring of anaerobic digestion in developing countries—a review. *Journal of Chemical Technology & Biotechnology*, 79(12), 1331-1341.

Lim, S.-J., Kim, B.J., Jeong, C.-M., Choi, J.-d.-r., Ahn, Y.H., Chang, H.N. 2008. Anaerobic organic acid production of food waste in once-a-day feeding and drawing-off bioreactor. *Bioresource Technology*, 99(16), 7866-7874.

Liu, D., Liu, D., Zeng, R.J., Angelidaki, I. 2006. Hydrogen and methane production from household solid waste in the two-stage fermentation process. *Water Research*, 40(11), 2230-2236.

Massé, D., Gilbert, Y., Topp, E. 2011. Pathogen removal in farm-scale psychrophilic anaerobic digesters processing swine manure. *Bioresource Technology*, 102(2), 641-646.

Mawioo, P.M., Rweyemamu, A., Garcia, H.A., Hooijmans, C.M., Brdjanovic, D. 2016. Evaluation of a microwave based reactor for the treatment of blackwater sludge. *Science of The Total Environment*, 548–549, 72-81.

Moosbrugger R. E., Wentzel M. C., Ekama G. A., Marais .GvR. 1993. Weak acid/bases and pH control in anaerobic systems-a review. *Water SA 19:1-10*.

Murto, M., Bjo"rnsson, L., Mattiasson, B. 2004. Impact of food industrial waste on anaerobic co-digestion of sewage sludge and pig manure. *Journal of Environmental Management* 70, 101-107.

Nallathambi Gunaseelan, V. 1997. Anaerobic digestion of biomass for methane production: A review. *Biomass and Bioenergy*, 13(1–2), 83-114.

Niwagaba, C., Kulabako, R.N., Mugala, P., Jönsson, H. 2009. Comparing microbial die-off in separately collected faeces with ash and sawdust additives. *Waste Management*, 29(7), 2214-2219.

Noike, T., Ko, I., Yokoyama, S., Kohno, Y., Li, Y. 2005. Continuous hydrogen production from organic waste. *Water Sci Technol*, 52(1-2), 145-51.

Olsen, J.E., Jørgensen, J.B., Nansen, P. 1985. On the reduction of Mycobacterium paratuberculosis in bovine slurry subjected to batch mesophilic or thermophilic anaerobic digestion. *Agricultural Wastes*, 13(4), 273-280.

Olsen, J.E., Larsen, H.E. 1987. Bacterial decimation times in anaerobic digestions of animal slurries. Institute of Hygiene and Microbiology, Royal Veterinary and Agricultural University, 13 Bülowsvej, DK-1870 Frederiksberg C, Denmark: http://www.researchgate.net/publication/223674948Bacterial_decimation_times_in_an aerobic_di gestions_of_animal_slurriess.

Pabón-Pereira, C.P., de Vries J.W., Slingerland M.A., G., Z., van Lier J.B. 2014. Impact of crop-manure ratios on energy production and fertilizing characteristics of liquid and solid digestate during co-digestion. *Environmental Technology*, 35(19), 2427-2434.

Pennington, T.H. 2001. Pathogens in agriculture and the environment. In: Pathogens in Agriculture and the Environment, Meeting organised by the SCI Agriculture and Environment Group, 16 October, SCI, London.

Rincón, B., Sánchez, E., Raposo, F., Borja, R., Travieso, L., Martín, M.A., Martín, A. 2008. Effect of the organic loading rate on the performance of anaerobic acidogenic fermentation of two-phase olive mill solid residue. *Waste Management*, 28(5), 870-877.

Riungu, J., Ronteltap, M., van Lier, J.B. 2018a. Build-up and impact of volatile fatty acids on *E. coli* and *A. lumbricoides* during co-digestion of urine diverting dehydrating toilet (UDDT-F) faeces. *J Environ Manage*, 215, 22-31.

Riungu, J., Ronteltap, M., van Lier, J.B. 2018b. Volatile fatty acids (VFA) build-up and its effect on *E. coli* inactivation during excreta stabilisation in single-stage and two-stage systems. *Journal of Water Sanitation and Hygiene for Development. 10.2166/washdev.2018.160*.

Romero-Güiza, M.S., Astals, S., Chimenos, J.M., Martínez, M., Mata-Alvarez, J. 2014. Improving anaerobic digestion of pig manure by adding in the same reactor a stabilizing agent formulated with low-grade magnesium oxide. *Biomass and Bioenergy*, 67, 243-251.

Sahlström, L. 2003. A review of survival of pathogenic bacteria in organic waste used in biogas plants. *Bioresource Technology*, 87(2), 161-166.

Santha, H., Sandino, J., Shimp, G.F., Sung, S. 2006. Performance Evaluation of a ‘Sequential-Batch& 8217; Temperature-Phased Anaerobic Digestion (TPAD) Scheme for Producing Class A Biosolids. *Water Environment Research*, 78(3), 221-226.

Schouten, M.A.C., Mathenge, R.W. 2010. Communal sanitation alternatives for slums: A case study of Kibera, Kenya. *Physics and Chemistry of the Earth, Parts A/B/C*, 35(13–14), 815-822.

Silva, S.A., Cavaleiro, A.J., Pereira, M.A., Stams, A.J.M., Alves, M.M., Sousa, D.Z. 2014. Long-term acclimation of anaerobic sludges for high-rate methanogenesis from LCFA. *Biomass and Bioenergy*, 67, 297-303.

Smith, S.R., Lang, N.L., Cheung, K.H.M., Spanoudaki, K. 2005. Factors controlling pathogen destruction during anaerobic digestion of biowastes. *Waste Management*, 25(4), 417-425.

Steffen, R., Szolar, O., Braun, R. 1998. Feedstocks for anaerobic digestion. Institute for Agrobiotechnology Tulln, University of Agricultural Sciences Vienna; 1998. Report no. 30-09.

Van Lier, J.B., Mahmoud, N., Zeeman, G. 2008. Biological wastewater treatment: Principles, modelling and design. Chapter 16: Anaerobic waste water treatment. IWA publishing.

Wang, K., Yin, J., Shen, D., Li, N. 2014a. Anaerobic digestion of food waste for volatile fatty acids (VFAs) production with different types of inoculum: Effect of pH. *Bioresource Technology*, 161, 395-401.

Wang, K., Yin, J., Shen, D., Li, N. 2014b. Anaerobic digestion of food waste for volatile fatty acids (VFAs) production with different types of inoculum: Effect of pH. *Bioresource Technology*, 161(0), 395-401.

Yuan, H., Chen, Y., Zhang, H., Jiang, S., Zhou, Q., Gu, G. 2006. Improved Bioproduction of Short-Chain Fatty Acids (SCFAs) from Excess Sludge under Alkaline Conditions. *Environmental Science & Technology*, 40(6), 2025-2029.

Zhang, B., He, P., LÜ, F., Shao, L. 2008. Enhancement of anaerobic biodegradability of flower stem wastes with vegetable wastes by co-hydrolysis. *Journal of Environmental Sciences*, 20(3), 297-303.

Zuo, Z., Wu, S., Zhang, W., Dong, R. 2014. Performance of two-stage vegetable waste anaerobic digestion depending on varying recirculation rates. *Bioresource Technology*, 162, 266-272.

Chapter 5: Anaerobic digestion of Urine Diverting Dehydrating Toilet Faeces (UDDT-FS) in urban poor settlements: Biochemical energy recovery

This Chapter is based on a paper: Riungu J., Ronteltap, M., van Lier, J.B. 2018. Anaerobic digestion of Urine Diverting Dehydrating Toilet Faeces (UDDT-FS) in urban poor settlements: Biochemical energy recovery; *Journal of Water Sanitation and Hygiene for Development* :WASHDev-D-18-00099.

Abstract

Biochemical energy recovery using digestion and co-digestion of faecal matter collected from urine diverting dehydrating toilet faeces (UDDT-FS) and mixed organic market waste (OMW) was studied under laboratory and pilot scale conditions. Laboratory scale biochemical methane potential (BMP) tests showed an increase in methane production with an increase in OMW fraction in the feed substrate. In subsequent pilot scale experiments, one-stage and two-stage plug flow digester were studied by, applying UDDT-FS:OMW ratios of 4:1 and 1:0, at about 10 and 12% Total Solids (TS) slurry concentrations. Comparable methane production was observed in one-stage ($R_{o-4:1, 12\%}$) (314 ± 15 mL CH_4/g VS added) and two-stage ($R_{am-4:1, 12\%}$) (325 ± 12 mL CH_4/g VS added) digesters, when applying 12% total solids (TS) slurry concentration. However, biogas production in $R_{am-4:1, 12\%}$ digester (571 ± 25 mL CH_4/g VS added) was about 12% higher than in $R_{o-4:1, 12\%}$, significantly more than the slight difference in methane production, i.e. 3–4%. The former was attributed to enhanced waste solubilisation and increased CO_2 dissolution, resulting from mixing the bicarbonate-rich methanogenic effluent for neutralisation purposes with the low pH (4.9) influent acquired from the pre-acidification stage. Moreover, higher process stability was observed in the first parts of the plug flow two-stage digester, characterised by lower VFA concentrations.

5.1 Introduction

As an innovative solution to enhance sanitary conditions in informal settlements in low income countries, Urine Diverting Dehydrating Toilets (UDDT) have been adopted (Austin & Cloete 2008; Niwagaba *et al.* 2009a; Schouten & Mathenge 2010; Katukiza *et al.* 2012). Such is also the approach adopted by Sanergy, a social enterprise working on sanitation improvement within LIHDS, in Nairobi, Kenya. Sanergy fabricates and installs the Fresh Life© toilets in collaboration with entrepreneurs in LIHDS who maintain them. Currently, approximately 7000 kg of faeces is collected from the UDDTs, further referred to as UDDT-FS, and delivered to a central treatment plant on a daily basis. Owing to the high pathogenic levels in human waste (Feachem *et al.* 1983), an extra pathogen inactivation step is required especially when the faecal matter will be valorised for agricultural purposes. A number of different treatment technologies were developed for source separated human faeces and include plain storage, composting, black soldier flies, chemical treatment, vermi-composting and anaerobic digestion (Vinnerås 2007; Niwagaba *et al.* 2009b; Rajagopal *et al.* 2013; Strande *et al.* 2014; Fagbohungbe *et al.* 2015). The main treatment technology applied by Sanergy for UDDT-FS is composting, producing an end product that is sold as organic manure (Evergrow@), available on the Kenyan market. Moreover, the increasing amount of collected UDDT-FS on a daily basis sparked a need for diversification of the treatment options.

The potential for application of anaerobic digestion (AD) at any scale and almost any place (Van Lier *et al.* 2008; Pabón-Pereira *et al.* 2014), marked the decision to select AD as a faecal waste treatment option in LIHDS. In addition, by means of AD, the chemically stored bio-energy in the organic waste can be recovered as biogas, providing an alternative fuel for local use (Abbasi *et al.* 2012). AD is considered an efficient technology for the stabilisation of organic wastes, producing a digestate with a high fertiliser value (Berndes *et al.* 2003; Martín-González *et al.* 2010; Park *et al.* 2016). The key reported drawback in AD is inadequate pathogen inactivation (Kunte *et al.* 2000; Chaggu 2004; Horan *et al.* 2004; Massé *et al.* 2011; Chen *et al.* 2012; Fagbohungbe *et al.* 2015) and low methane production especially from human faecal matter (Rajagopal *et al.* 2013; Fagbohungbe *et al.* 2015). It must be noted that the microbiological safety of the digestate and treated sludge is essential as it has implications for human health and cycling of pathogens in a densely populated environment through the food chain (Avery *et al.* 2014). As such, this study is part of a wider research on the potentials for the anaerobic stabilisation of UDDT-FS, enhancing biogas production and pathogen inactivation, with the present paper focusing on the production of another side-product next to hygienised sludge, i.e. biogas.

In Chapter 4, we evaluated the accumulation of volatile fatty acids and their effect on pathogen inactivation during the digestion of UDDT-FS and mixtures of UDDT-FS and organic market waste (OMW) in a one- and two-stage plug flow anaerobic digester (Riungu *et al.* 2018b). Results showed higher pathogen inactivation in the two-stage plug flow digester, with the digestate meeting WHO standards of 1000 CFU/100 mL, applying a solids retention time

(SRT) of 29 days. The used OMW, widely available and at close proximity to UDDT-FS source, is characterised by a vast readily degradable organic fraction (Zhang *et al.* 2008; Riungu *et al.* 2018a). In addition to pathogen removal, the production of an alternative fuel (biogas) from the faecal matter will very likely accelerate the acceptance of the proposed technology. As such, the here described research focused on the potential for biogas production during anaerobic stabilisation of UDDT-FS using laboratory-scale BMP tests and pilot-scale plug flow one- and two-stage anaerobic digesters. Under pilot scale experiments one-stage and two-stage plug flow digester were researched, applying UDDT-FS:OMW ratios of 4:1 and 1:0, at about 10 and 12% Total Solids (TS) slurry concentrations.

5.2 Materials and Methods

5.2.1 UDDT-FS waste samples

UDDT-FS samples used for this study were obtained from the Fresh Life® UDDT within Mukuru Kwa Njenga/ Mukuru Kwa Reuben LIHDS, Kenya. Fresh Life@ toilets are offered on a pay-and-use basis in the form of serviced shared facilities, charging between 0.05–0.1 euros per use. Within each toilet facility, a 30 L container is used for waste collection, with approximately 10 g sawdust added by the user after every toilet use. The toilets are emptied on a daily basis, where used containers are replaced by clean ones. Five containers with UDDT-FS were randomly selected after which mixing of the contents was done in order to obtain a homogeneous mix.

5.2.2 Organic market waste samples

OMW was collected from vegetable vendors, eating points and waste disposal points within Mukuru Kwa Njenga and Mukuru Kwa Reuben LIHDS. About 20 kg of the waste was collected and contained food waste, vegetable waste and fruit waste, in equal proportions. Size reduction was achieved by manual chopping to about 1 cm size for pilot scale test substrates whereas samples for laboratory scale tests were blended using Ramton© domestic blender for 1 minute. Table 5.1 shows the characteristics of the UDDT-FS and OMW that was used in the study. After collection the waste was refrigerated at 4 °C to minimise bioconversion of the samples prior to testing.

Table 5.1: Characterisation of urine diverting dehydrating toilets waste and mixed organic market waste used in study (Riungu et al. 2018a)

		UDDT-FS		OMW	
		Value	STDEV	Value	STDEV
Total solids (TS)	(%)	24.5	3.8	17.9	1.6
Moisture content	(%)	75.5	3.8	80.7	4.1
Volatile solids (VS)	(% wgt)	20.1	3.5	16.9	4.4
Escherichia coli (E. coli)	(CFU/g TS)	1.7E+09	5.3E+08	2.7E+05	7.4E+04
Ascaris eggs		Not detected		Not detected	

5.2.3   Inoculum

Inoculum for the anaerobic digestion experiments used in this study was obtained from an onsite fixed dome anaerobic digester within Kibera informal settlement, Kenya. The bio-centre was erected by Umande Trust, a non-governmental organisation (https://umande.org/) and managed in partnership with a community-based organisation, Kibera Kids Youth Organisation (KIDYOT). The inoculum upon collection was incubated for 1 week to methanise any organic matter before use.

5.2.4   Experimental setup

5.2.4.1 Laboratory scale BMP test

Biochemical methane potential (BMP) test experiments were performed to access methane production during anaerobic stabilisation of faecal waste. Three substrate ratios, based on our previous study (Riungu *et al.* 2018a), that investigated the effect of volatile fatty acids (VFAs) on pathogen inactivation were applied; UDDT-FS:OMW ratios 1:0, 4:1 and 0:1. An inoculum to substrate ratio of 2:1 (Zeng *et al.* 2010) was used, maintaining approximately 1.5 g volatile solids (VS)/100 mL solution, based on initial VS concentration of inoculum and substrate.

Batch digestion experiments were conducted in triplicate using 100 mL glass serum vials (80 mL working volume). After adding the required amounts of substrate and inoculum in each serum vial, basic anaerobic medium (BAM) was added according to Angelidaki *et al.* (2009) (Table 5.2), in addition to 1 g/L sodium carbonate buffer. Hereafter, tap water was added to a volume of 80 mL. The vials were sealed with butyl rubber stoppers and flushed with argon gas for 30 seconds to purge out oxygen. The vials were incubated at 35($\pm$1) °C for 30 days, with manual mixing. Triplicate blanks that contained inoculum and BAM were incubated in order to correct for gas production from the inoculum. Gas pressure in the digesters was measured regularly with a digital pressure meter model GMH 3150 (Greisinger, Germany) utilising a sensor model MSD 4 BAE with a resolution of 1 mbar.

Table 5.2 Nutrients applied for BMP test

	Composition (g/L)	Dose (ml/L)		Composition (g/L)	Dose (ml/L)
Macronutrients					
NH$_4$Cl	170	2			
KH$_2$PO$_4$	37	2			
CaCl$_2$2H$_2$O	8	2			
MgSO$_4$.4H$_2$O	9	2			
Trace elements and micro nutrients					
FeCl$_3$.4H$_2$O	2	1	Resazurine	0.5	1
ZnCl$_2$	0.05	1	HCl (36%)	1 mL/L	1
H$_3$BO$_3$	0.05	1	EDTA	1	1
CuCl$_2$.2H$_2$O	0.03	1	NiCl$_2$.6H$_2$O	0.05	1
MnCl$_2$.4H$_2$O	0.5	1	Na$_2$SeO$_3$.5H$_2$O	0.1	1
CoCl$_2$.6H$_2$O	2	1	Yeast extract	0.1	1
(NH$_4$)6Mo$_7$O$_{24}$.4H$_2$O	0.09	1	NaC$_2$H$_3$O$_2$-3H$_2$O	1g COD/L	1g COD/L

5.2.4.2 Pilot scale AD experiments

Pilot scale substrate selection was based on a series of laboratory scale batch-tests derived from previous experimental data (Riungu *et al.* 2018a) applying UDDT-FS:OMW ratios of 4:1 and 1:0. In addition, research aimed at treating the highest possible substrate's TS concentration that can freely flow through the plug flow digester without the necessity of using pumps. As such, 12% TS was chosen as the highest substrate TS concentration with additional experiments at 10% TS for assessing the impact of lower TS concentrations on biogas production.

Two sets of digesters were used, namely a one-stage digester (R$_o$) and a two-stage digester (R$_{am}$) comprising a hydrolysis/ acidogenic digester (R$_a$) and a methanogenic digester (R$_m$).

Hydrolysis digester

The hydrolysis digesters (R_a) were fabricated from 30 L plastic containers, with a working volume of 20 L. These digesters were equipped with a cover, incorporated with two separate ports, i.e. a feeding port and a port fixed with a manual stirring mechanism, whereas the bottom of each digester was equipped with a discharge/ effluent valve.

Plug flow digester

Six plug flow digesters (Figure 5.1) were constructed using 175 L tubular polyethylene bags, with polyethylene material thickness being 0.2 mm. The digesters had a liquid capacity of 145 L, with up to 30 L available for in-vessel biogas storage. The majority of biogas produced flowed by pressure to a 175 L biogas storage bag that was installed directly above each digester. In addition, three separate ports were incorporated onto each bag: inlet port (SP_1); sampling port (SP_2) at 0.7 m digester length; a gas discharge port at 1.4 m digester length; and effluent/discharge port (SP_3) at 2.1 m digester length. A total solids retention time (SRT) of 29 days was maintained for the anaerobic digestion process.

Figure 5.1 Plug flow digester layout; digesters on the floor, biogas collection bags directly above; sampling points at different length of the digester are indicated as SP_1, SP_2 and SP_3, respectively.

5.2.5 Plug flow digester start-up and operation in one- and two-stage AD

Digesters were inoculated using the inoculum described above under "Inoculum". The six plug flow digesters D_1, D_2, D_3, D_4, D_5 and D_6, were divided into two groups, D_1-D_3-D_5 and D_2-D_4-D_6, referring to one-stage digestion of UDDT-FS:OMW ratio 1:0 at 12% TS ($R_{o-1:0,\ 12\%}$) and UDDT-FS:OMW ratio 1:0 at 10% TS ($R_{o-1:0,\ 10\%}$), respectively. Every morning, 5 L/day of the substrate was fed to the respective digesters. Stabilisation of the digesters was achieved after 6 weeks, and sample collection and analysis commenced and continued for a further 8 weeks.

The impact of co-digestion on biogas production and organic matter stabilisation was assessed by applying both one- and two-stage digesters, utilising a UDDT-FS:OMW ratio of 4:1 at 12% TS, i.e. $R_{o-4:1, 12\%}$ and $R_{am-4:1, 12\%}$, respectively. In these experiments, the six plug flow digesters were also divided into two treatments groups, where digesters D_1-D_3-D_5 consisted of the two-stage $R_{am-4:1, 12\%}$ digesters and digesters D_2-D_4-D_6 comprised the one-stage $R_{o-4:1, 12\%}$ digesters. Every morning, 5 L of feed substrate was fed into the one- and two-stage digesters, with feed substrate being prepared as follows: 1) One-stage: freshly prepared UDDT-FS:OMW ratio of 4:1 at 12% TS concentration, 2) Two-stage: Hydrolysis/acidogenic (R_a) digester effluent acted as influent to the methanogenic digesters (R_m). The pH of R_a effluent (4.9±0.1) was adjusted by titration using two-stage (R_{am}) digester effluent to a range of 5.8–6.2 prior to feeding it to the R_m digesters. For all digesters, stabilisation of biogas production was achieved after two months when data collection commenced. Finally, the concentration of the feed into $R_{o-4:1, 12\%}$ was reduced to 10% TS.

Samples from experiments were taken on a weekly basis for analysis of TS and volatile solids (VS), whereas biogas and methane analysis were carried out on a daily basis over the entire experimental period.

5.2.6 Analytical procedures

Biogas production in laboratory scale BMP vials was determined by measuring the pressure increase in the headspace volume (20 mL) using a digital pressure meter model GMH 3150 (Greisinger, Germany) utilising a sensor model MSD 4 BAE with a resolution of 1 mbar. The volumetric biogas production was calculated from the assessed pressure increase and expressed under standard temperature and pressure (STP, 0 °C and 760 mm Hg) according to the following equation (Pabon Pereira *et al.* 2012):

$$V_{Biogas} = \frac{P.V_h V_{mol}}{R.T} \qquad (1)$$

where P is biogas pressure in the vial (kPa); V_h is digester headspace volume (L); V_{mol} is molar gas volume at 308 K (L/mol); R is the universal gas constant (8.31 kPa L/mol K) and T is temperature (K).

The net gas production for calculating the BMP values was obtained by subtracting the gas production of the blank samples.

Biogas flow measurements in the pilot scale digesters were performed using American Meter Company gas flow meters (Model AC-250) with IMAC Systems pulse digital counters and a vacuum pump.

Determination of methane content in biogas in laboratory and pilot scale experiments was performed by liquid displacement method. Herein, a known amount of biogas was passed through a 5% sodium hydroxide solution to strip CO_2. Under laboratory scale BMP test, methane measurement was carried out twice a week while in pilot scale test, methane

measurement was done once a day. In this approach the quantity of H_2S in the biogas is
considered negligible.

The percentage methane fraction in biogas was obtained by:

$$\%CH_4 = \frac{Volume\ of\ displaced\ NaOH\ solution}{Volume\ of\ gas\ injected} * 100 \qquad (2)$$

Methane production was then calculated by multiplying the mean corrected biogas
volume produced in a specified time lapse by the assessed average percentage methane content
in the biogas, whereas methane yields were obtained by dividing the total methane volume
produced in the specified time lapse by the weight of the substrate (VS_{added} (in g)) fed to the
plug flow digesters in the same time lapse, according to the following equation:

$$V_{CH_4} = \%CH_4 \frac{V'\ biogas - V''\ biogas}{VS} \qquad (3)$$

where % CH_4: fraction of methane in biogas; V' biogas is the volume of biogas produced
on the substrate; V'' biogas is the volume of biogas produced by the blank; and VS is volatile
solids added (g)

Total solids (TS) and volatile solids (VS) analysis were conducted according to the
gravimetric method (SM-2540D and SM-2540E), as outlined in the Standard Methods for the
Examination of Water and Wastewater (APHA 1995).

5.2.7   Data analysis

Bivariate Pearson's correlation test was used to assess trends in methane production
from individual digesters within a given experiment. From each of the three trials, the data
obtained was analysed by computing the averages, standard deviations and standard errors.
Results obtained were presented either in table or Figure form.

5.3 Results and discussion

5.3.1 Methane production in batch scale BMP tests

Figure 5.2 shows cumulative methane produced against time for UDDT-FS:OMW
ratios 1:0, 4:1 and 0:1. Highest methane production was recorded within the first 10 days of the
experiment, with UDDT-FS:OMW ratio 0:1 attaining 45.8 mL CH_4/g VS added/day (Figure
5.2) whereas UDDT-FS:OMW ratios 1:0 and 4:1 depicted 27.3 and 17.1 mL CH_4/g VS
added/day respectively. After the 10th day, a decline in methane production was observed up
to the 30th day of the experiment.

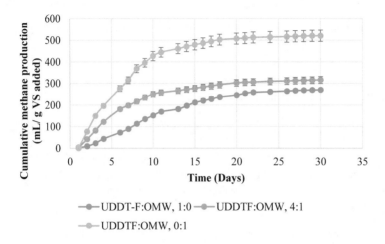

Figure 5.2 Cumulative methane production against time in anaerobic digestion of UDDT-FS ratios 1:0, 4:1 and 0:1.

Overall, 271±13, 315±26 and 521±36 mL CH$_4$/ g VS$_{added}$ was recorded from UDDT-FS:OMW ratios 1:0, 4:1 and 0:1 respectively (Figure 5.2). An average of about 0.26–0.30 L CH$_4$/g VS$_{added}$ has been reported in batch scale BMP assays of black water (Rajagopal *et al.* 2013), and about 250 mL CH$_4$/g VS$_{added}$ in anaerobic digestion of human faecal material (faeces + urine) (Fagbohungbe *et al.* 2015). The findings showed an increasing trend in biogas production with the increase in OMW fraction within the feed substrate, which is congruent to the observed higher VFA build-up at increasing OMW fractions in our previous work (Riungu *et al.* 2018a). The produced VFA was subsequently converted to biogas. In practical situations where biogas generation is the main driver for implementing AD, the use of OMW as sole substrate may lead to excessive VFA build-up and subsequent system acidification (Angeriz-Campoy *et al.* 2015; Riungu *et al.* 2018a). In our previous work, pH levels declined to below 4 at UDDT-FS:OMW ratios lower than 1:2, inhibiting methanogenic activity. In general, OMW is carbohydrate rich, has a high C/N ratio and is easily hydrolysable (Gómez *et al.* 2006; Lim *et al.* 2008), in addition to containing appreciable amounts of fats that are easily hydrolysable to long chain fatty acids (Silva *et al.* 2014; Angeriz-Campoy *et al.* 2015). As such, in co-digestion of OMW and UDDT-FS, both substrates complement each other: UDDT-FS is characterised by a low carbon to nitrogen ratio (Mata-Alvarez *et al.* 2011; Fonoll *et al.* 2015) and low methane production (Rajagopal *et al.* 2013), it provides adequate micro/macro nutrients, alkalinity and moisture content (Silvestre *et al.* 2015).

5.3.2 Methane production pilot scale experiments

The experiments evaluated the impact of digester configuration, co-digestion and substrate concentration on the accumulating methane production during an 8-week time period. All digesters showed a linear increase in accumulating methane with time (Figure 5.3). Overall, the obtained trend in accumulated methane production per g VS added was in the order: $R_{am-4:1, 12\%}$> $R_{am-4:1, 10\%}$>$R_{o-1:0, 12\%}$> $R_{o-1:0, 12\%}$> $R_{o-1:0, 10\%}$, with minimal differences between the co-digesting experiments (Figure 5.5). Results from bivariate Pearson's correlation test performed on triplicate samples within each experiment showed high and significant correlation in methane production within a particular experiment, all being within the range of $r=0.474^{**}$–0.840^{**} (**correlation is significant at the 0.01 level (2-tailed)), indicating good digester progress throughout the experimental period.

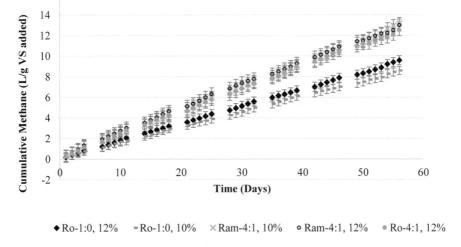

♦ Ro-1:0, 12%   ▪ Ro-1:0, 10%   ✕ Ram-4:1, 10%   ○ Ram-4:1, 12%   ● Ro-4:1, 12%

Figure 5.3 | Cumulative methane produced (L/g VS added) against time (days)

The effect of digester configuration was gauged by applying a UDDT: OMW ratio of 4:1 at 12% TS in a one-stage ($R_{o-4:1, 12\%}$) and two-stage digester ($R_{am-4:1, 12\%}$). Methane production in $R_{o-4:1, 12\%}$ and $R_{am-4:1, 12\%}$ digester was comparable with corresponding values being 314±15 and 325±12 mL CH_4 /g VS added (Figure 5.5), respectively.

However, average biogas production in $R_{am-4:1, 12\%}$ was 571±25 mL CH_4/g VS added and was about 12% higher than in the $R_{o-4:1, 12\%}$ system which is significantly more than the slight difference in methane production, i.e. 3–4%. The small difference in methane production may be attributed to enhanced waste solubilisation in the two-stage digester, as reported in our previous study (Riungu *et al.* 2018b) and in agreement with related studies (Zuo *et al.* 2014; De Gioannis *et al.* 2017; Gaby *et al.* 2017). On the other hand, the observed higher biogas production in the two-stage digester likely can be ascribed to increased CO_2 dissolution, resulting from mixing the bicarbonate-rich methanogenic effluent for neutralisation purposes with the low pH (4.9) influent coming from the pre-acidification stage. The latter also explains the lower CH_4 content in the produced biogas in the gas bags of the two-stage digester and the higher pH in the effluent (Table 5.3). In the two-stage set-up, part of the produced acidity is already lost as CO_2 in the pre-acidification step that was open to air, leading to a higher overall alkalinity of the methanogenic effluent compared to the one-stage process.

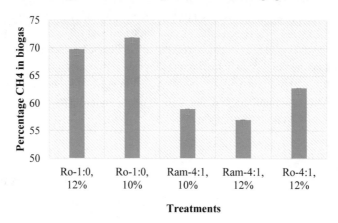

Figure 5.4 Percentage methane content in biogas during anaerobic stabilisation of UDDT-FS for; two-stage digester ($R_{am-4:1, 12\%}$ and $R_{am-4:1, 10\%}$) and one-stage digester ($R_{o-4:1, 12\%}$, $R_{o-1:0, 12\%}$ and $R_{o-1:0, 12\%}$).

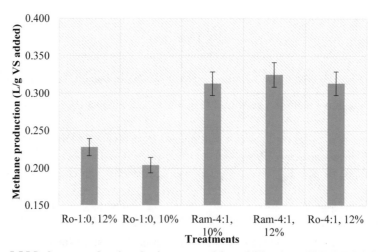

Figure 5.5 Methane production during anaerobic stabilisation of UDDT-FS for; two-stage digester ($R_{am-4:1, 12\%}$ and $R_{am-4:1, 10\%}$) and one-stage digester ($R_{o-4:1, 12\%}$, $R_{o-1:0, 12\%}$ and $R_{o-1:0, 10\%}$).

In the two-stage digestion set-up with digestate recycling, chances for possible acidification in the front part of the methanogenic plug flow digester is reduced. An active methanogenic activity in the front part of $R_{am-4:1, 12\%}$ digester as indicated by a decline in total volatile acids (TVFA) concentrations between SP_1 and SP_3, and their subsequent conversion to biogas (Table 5.3) was observed. As mentioned before, the stabilised methanogenic conditions in the early stages of the plug flow digester of the two-stage set-up were achieved by digestate or effluent recycling that re-introduces active methanogenic biomass upfront (Cavinato et al. 2011). However, the $R_{o-4:1, 12\%}$ digester showed an increasing trend in total volatile acids (TVFA), and non-dissociated volatile acids (ND-VFA) build-up between SP_1 and SP_2, resulting in an acidic pH 5.4 thus indicating predominating acidogenesis in the first part of the plug flow digester (Table 5.3). In this digester, highest methanogenic activity was observed between SP_2 and SP_3, where high reduction in TVFA indicated their subsequent conversion to biogas.

The impact of co-digestion on methane production in the one-stage digestion process was assessed by applying a UDDT:OMW ratio of 4:1, at 12% TS ($R_{o-4:1, 12\%}$) and a UDDT:OMW ratio of 1:0, at 12% TS ($R_{o-1:0, 12\%}$). Methane production in $R_{o-4:1, 12\%}$ and $R_{o-1:0, 12\%}$ digester system was 314±15 and 228±19 l CH_4/g VS added (Figure 5.5) respectively, representing a 37% increase when OMW was added. The corresponding percentage methane content in biogas in $R_{o-4:1, 12\%}$ and $R_{o-1:0, 12\%}$ digester system was 62.8±2 and 70.0±4.5% respectively (Figure 5.4). The lower CH_4 content in the biogas of the co-digester comes from the OMW fraction of the feed substrate which is generally characterised by carbohydrate rich organic matter with a somewhat higher oxidation state than the UDDT-FS, which agrees with the lower COD/TOC ratio for OMW as presented in Table 5.1. Also, the higher methane production (in L CH_4/g VS added) in $R_{o-4:1, 12\%}$ is attributable to the highly digestible OMW fraction, reflected by the high TVFA build-up attained during the digestion process (Table 5.2). Intrisically, during co-digestion, hydrolysis of OMW enhances TVFA build-up in the digestion medium (Zhang et al. 2005; Zhang et al. 2008; Riungu et al. 2018a), which is subsequently converted to biogas but may lead to subsequent system acidification (Angeriz-Campoy et al. 2015) if used as sole substrate.

Table 5.3 Variation in VFA and pH along the digester length (adopted from Riungu et al. 2018b)

	Digester	Parameter	SP$_1$	SP$_2$	SP$_3$
Co-digestion UDDT-FS:OMW ratio 4:1	R$_{m-4:1, 12\%}$	TVFA (mg/L)	15685±1772	10526±844	1575±607
		ND-VFA (mg/L)	800±112	286±68	1.7±0.2
		ND-VFA (%)	5.1±0.6	2.7±0.6	0.1
		pH	6.4 ±0.1	6.4±0.1	7.8±0.1
	R$_{m-4:1, 10\%}$	TVFA (mg/L)	12347±887	8702±72	1744±101
		ND-VFA (mg/L)	660±311	281±49	1.6±0.3
		ND-VFA (%)	3.5±2	3.2±0.6	0.1
		pH	6.3±0.1	6.2±0.1	7.8±0.1
	R$_{o-4:1,12\%}$	TVFA (mg/L)	3844±679	12121±1153	2629±326
		ND-VFA (mg/L)	599.4±150	2379±409	5±1.2
		ND-VFA (%)	15.8±3.4	19.6±2.8	0.2
		pH	5.4±0.1	5.4±0.1	7.5±0.1

The potential impact of substrate concentration on methane production was investigated applying two substrates concentrations, i.e. 12% and 10% TS. These concentrations were applied in both the one-stage and two-stage plug flow digesters, at UDDT-FS:OMW ratios of 1:0 and 4:1 respectively. Using both digester systems, slightly higher methane production was observed at 12% TS than 10% TS. The two-stage digesters R$_{am-4:1,12\%}$ produced 325±12 mL CH$_4$/g VS$_{added}$ whereas the corresponding value in R$_{am-4:1,10\%}$ was 313±17 mL CH$_4$/g VS$_{added}$. Similarly, in the one-stage digester, R$_{o-1:0,12\%}$ and R$_{o-1:0,10\%}$ the observed methane production was 228±19 mL CH$_4$/g VS and 204±22 mL CH$_4$/g VS$_{added}$ respectively. The percentage methane in the biogas of R$_{am-4:1,12\%}$ and R$_{am-4:1,10\%}$ digesters at 10 and 12% TS were 57±8 and 59±4%, respectively (Figure 5.3). Corresponding values in R$_{o-1:0,12\%}$ and R$_{o-1:0,10\%}$ digesters being 70±2 and 71±8% respectively. Furthermore, all digesters showed stable digestion performance with effluent pH in the one-stage digesters a bit lower compared to the two-stage digesters, i.e. 7.3–7.7 and 7.6–8.1, respectively. The average pH in the methanogenic stage is considered optimal for the methanogenic biomass, i.e. 7.5–8.1.

5.3.3 AD application of UDDT-FS management in LIHDS

This study is part of a wide research seeking to enhance biogas production and pathogen inactivation from UDDT-FS in LIHDS. The key objective of the research was to maximise the amounts of UDDT-FS that can be treated, while producing biogas and stabilised digestate that can be used for agricultural applications. The findings obtained in this research demonstrated the technical feasibility of AD technology in UDDT-FS management. Moreover, in addition to efficient management of the waste, the produced biogas has a wide range of applications.

The possibilities of reactor failure are apparent during co-digestion, especially at increased OMW fraction due to reactor acidification. The UDDT-FS:OMW ratio 4:1 adopted in this study was based on recommendations from our previous study (Riungu et al. 2018a). Increasing the OMW fraction in the feed substrate leads to rapid acidification thereby lowering the pH and increasing the ND-VFA concentration that has a toxic effect not only to pathogens but all anaerobic bacterial population, thus process failure. As such, precaution should be taken to ensure application of optimal UDDT-FS:OMW ratios during co-digestion.

Results showed that co-digestion in the proposed plug-flow digester produced a low pathogen content–digestate, i.e. $<1*10^3$ CFU/100 mL (Riungu et al. 2018b)) and a biogas stream that can be used as an alternative fuel source for LIHDS residents, delivering about 6500 MJ/month for a bio-centre with a user load of 500 persons/day. The system presents a cost-effective solution for the many LIHDS in sub-Saharan Africa: the plug-flow reactor can be assembled with locally available materials and the high population density assures a constant supply of raw materials, whereas the prevailing high temperatures ensures the system's zero energy operating requirements.

5.4 Conclusions

Experiments were conducted to investigate the biochemical energy recovery during digestion and co-digestion of faecal matter collected from urine diverting dehydrating toilet faeces (UDDT-FS) and mixed organic market waste (OMW) under laboratory and pilot scale conditions. Laboratory scale biochemical methane potential (BMP) tests showed a positive correlation between methane production and increasing OMW fraction in the feed substrate.

Under pilot scale conditions, comparable methane production was observed in one-stage ($R_{o-4:1, 12\%}$) (314 ± 15 mL CH_4/g VS added) and two-stage ($R_{am-4:1, 12\%}$) (325 ± 12 mL CH_4/g VS added) digesters, when applying 12% total solids (TS) slurry concentration. However, biogas production in $R_{am-4:1, 12\%}$ digester (571 ± 25 mL CH_4/g VS added) was about 12% higher than in the $R_{o-4:1, 12\%}$, significantly more than the slight difference in methane production, i.e. 3–4%. The increased methane and biogas production was attributed to enhanced waste solubilisation and increased CO_2 dissolution, resulting from mixing the bicarbonate-rich methanogenic effluent for neutralisation purposes with the low pH (4.9) influent coming from the pre-acidification stage. Moreover, compared to the one-stage reactor, higher process stability was observed in the first parts of the two-stage plug flow digester, characterised by lower VFA concentrations. The observed high VFA concentrations and acidic pH (5.4) in the first parts of one-stage digester indicate low process stability, particularly with increased OMW fractions in the feed substrate.

Within the wide research, overall findings have shown the potential application of two-stage AD technology in addressing the human waste menace, especially in LIHDS. The proposed system can be applied at either small scale or large scale, depending on available space. The treatment system has almost zero energy requirements when implemented in warm areas where optimal mesophilic temperatures can be reached without heating.

Acknowledgements

This research is funded by the Bill & Melinda Gates Foundation under the framework of SaniUp project (Stimulating local Innovation on Sanitation for the Urban Poor in Sub-Saharan Africa and South-East Asia) (OPP1029019). The authors would like to thank Ani Vabharneni, Sanergy Kenya, and DVC-ARS, Meru University, Kenya for their valuable support during this study.

References

Abbasi, T., Tauseef, S.M., Abbasi, S.A. 2012. Anaerobic digestion for global warming control and energy generation - an overview. Renew. Sust. Energy Rev. 16, 3228-3242.

Angeriz-Campoy, R., Álvarez-Gallego, C.J., Romero-García, L.I. 2015. Thermophilic anaerobic co-digestion of organic fraction of municipal solid waste (OFMSW) with food waste (FW): Enhancement of bio-hydrogen production. *Bioresource Technology*, 194, 291-296.

APHA. 1995. Standard Methods for the Examination of Water and Wastewater, nineteenth ed. American Public Health Association, Washington DC, USA.

Austin, L.M., Cloete, T.E. 2008. Safety aspects of handling and using fecal material from urine-diversion toilets - a field investigation. *Water Environ. Res.*, 80(4), 308-315.

Avery, L.M., Anchang, K.Y., Tumwesige, V., Strachan, N., Goude, P.J. 2014. Potential for Pathogen reduction in anaerobic digestion and biogas generation in Sub-Saharan Africa. *Biomass and Bioenergy*, 70(0), 112-124.

Berndes, G., Hoogwijk, M., van den Broek, R., . . 2003. The contribution of biomass in the future global energy supply: a review of 17 studies. Biomass. Bioeng. 25, 1e28.

Cavinato, C., Bolzonella, D., Fatone, F., Cecchi, F., Pavan, P. 2011. Optimization of two-phase thermophilic anaerobic digestion of biowaste for hydrogen and methane production through reject water recirculation. *Bioresource Technology*, 102(18), 8605-8611.

Chaggu, E.J. 2004. Sustainable Environmental Protection Using Modified Pit-Latrines. Ph.D Thesis, Wageningen University, The Netherlands.

Chen, Y., Fu, B., Wang, Y., Jiang, Q., Liu, H. 2012. Reactor performance and bacterial pathogen removal in response to sludge retention time in a mesophilic anaerobic digester treating sewage sludge. *Bioresource Technology*, 106(Supplement C), 20-26.

De Gioannis, G., Muntoni, A., Polettini, A., Pomi, R., Spiga, D. 2017. Energy recovery from one- and two-stage anaerobic digestion of food waste. *Waste Management*, 68, 595-602.

Fagbohungbe, M.O., Herbert, B.M.J., Li, H., Ricketts, L., Semple, K.T. 2015. The effect of substrate to inoculum ratios on the anaerobic digestion of human faecal material. *Environmental Technology & Innovation*, 3(0), 121-129.

Feachem, R.G., Bradley, D.J., Garelick, H., D., M.D. 1983. Sanitation and Disease Health Aspects of Excreta and Wastewater Management. *Report No.:11616 Type: (PUB)*

Fonoll, X., Astals, S., Dosta, J., Mata-Alvarez, J. 2015. Anaerobic co-digestion of sewage sludge and fruit wastes: Evaluation of the transitory states when the co-substrate is changed. *Chemical Engineering Journal*, 262(0), 1268-1274.

Gaby, J.C., Zamanzadeh, M., Horn, S.J. 2017. The effect of temperature and retention time on methane production and microbial community composition in staged anaerobic digesters fed with food waste. *Biotechnology for Biofuels*, 10(1), 302.

Gómez, X., Morán, A., Cuetos, M.J., Sánchez, M.E. 2006. The production of hydrogen by dark fermentation of municipal solid wastes and slaughterhouse waste: A two-phase process. *Journal of Power Sources*, 157(2), 727-732.

Horan, N.J., Fletcher, L., Betmal, S.M., Wilks, S.A., Keevil, C.W. 2004. Die-off of enteric bacterial pathogens during mesophilic anaerobic digestion. *Water Research*, 38(5), 1113-1120.

Katukiza, A.Y., Ronteltap, M., Niwagaba, C.B., Foppen, J.W.A., Kansiime, F., LensP.N.L. 2012. Sustainable sanitation technology options for urban slums, *Biotechnology Advances*, Article in press.

Kunte, D.P., Yeole, T.Y., Ranade, D.R. 2000. Inactivation of Vibrio cholerae during anaerobic digestion of human night soil. *Bioresource Technology*, 75(2), 149-151.

Lim, S.-J., Kim, B.J., Jeong, C.-M., Choi, J.-d.-r., Ahn, Y.H., Chang, H.N. 2008. Anaerobic organic acid production of food waste in once-a-day feeding and drawing-off bioreactor. *Bioresource Technology*, 99(16), 7866-7874.

Martín-González, L., Colturato, L.F., Font, X., Vicent, T. 2010. Anaerobic co-digestion of the organic fraction of municipal solid waste with FOG waste from a sewage treatment plant: Recovering a wasted methane potential and enhancing the biogas yield. *Waste Management*, 30(10), 1854-1859.

Massé, D., Gilbert, Y., Topp, E. 2011. Pathogen removal in farm-scale psychrophilic anaerobic digesters processing swine manure. *Bioresource Technology*, 102(2), 641-646.

Mata-Alvarez, J., Dosta, J.S.M., S.,, Astals, S. 2011. Codigestion of solid wastes: a review of its uses and perspectives including modeling. *Crit. Rev. Biotechnol.*, 31, 99-111.

Niwagaba, C., Kulabako, R.N., Mugala, P., Jönsson, H. 2009. Comparing microbial die-off in separately collected faeces with ash and sawdust additives. *Waste Management*, 29(7), 2214-2219.

Pabón-Pereira, C.P., de Vries J.W., Slingerland M.A., G., Z., van Lier J.B. 2014. Impact of crop-manure ratios on energy production and fertilizing characteristics of liquid and solid digestate during co-digestion, *Environmental Technology*, 35(19), 2427-2434.

Pabon Pereira, C.P., Castanares, G., van Lier, J.B. 2012. An OxiTop((R)) protocol for screening plant material for its biochemical methane potential (BMP). *Water Sci Technol*, 66(7), 1416-23.

Park, K.Y., Jang, H.M., Park, M.-R., Lee, K., Kim, D., Kim, Y.M. 2016. Combination of different substrates to improve anaerobic digestion of sewage sludge in a wastewater treatment plant. *International Biodeterioration & Biodegradation*, 109, 73-77.

Rajagopal, R., Lim, J.W., Mao, Y., Chen, C.-L., Wang, J.-Y. 2013. Anaerobic co-digestion of source segregated brown water (feces-without-urine) and food waste: For Singapore context. *Science of The Total Environment*, 443, 877-886.

Riungu, J., Ronteltap, M., van Lier, J.B. 2018a. Build-up and impact of volatile fatty acids on *E. coli* and *A. lumbricoides* during co-digestion of urine diverting dehydrating toilet (UDDT-F) faeces. *J Environ Manage*, 215, 22-31.

Riungu, J., Ronteltap, M., van Lier, J.B. 2018b. Volatile fatty acids (VFA) build-up and its effect on *E. coli* inactivation during excreta stabilisation in single-stage and two-stage systems. *Journal of Water Sanitation and Hygiene for Development*. *10.2166/washdev.2018.160*.

Schouten, M.A.C., Mathenge, R.W. 2010. Communal sanitation alternatives for slums: A case study of Kibera, Kenya. *Physics and Chemistry of the Earth, Parts A/B/C*, 35(13–14), 815-822.

Silva, S.A., Cavaleiro, A.J., Pereira, M.A., Stams, A.J.M., Alves, M.M., Sousa, D.Z. 2014. Long-term acclimation of anaerobic sludges for high-rate methanogenesis from LCFA. *Biomass and Bioenergy*, 67, 297-303.

Silvestre, G., Bonmatí, A., Fernández, B. 2015. Optimisation of sewage sludge anaerobic digestion through co-digestion with OFMSW: Effect of collection system and particle size. *Waste Management*, 43, 137-143.

Van Lier, J.B., Mahmoud, N., Zeeman, G. 2008. Biological wastewater treatment: Principles, modelling and design. Chapter 16: Anaerobic waste water treatment. IWA publishing.

Zeng, S., Yuan, X., Shi, X., Qiu, Y. 2010. Effect of inoculum/substrate ratio on methane yield and orthophosphate release from anaerobic digestion of Microcystis spp. *Journal of Hazardous Materials*, 178(1), 89-93.

Zhang, B., He, P., LÜ, F., Shao, L. 2008. Enhancement of anaerobic biodegradability of flower stem wastes with vegetable wastes by co-hydrolysis. *Journal of Environmental Sciences*, 20(3), 297-303.

Zhang, B., Zhang, L.L., Zhang, S.C., Shi, H.Z., Cai, W.M. 2005. The influence of pH on hydrolysis and acidogenesis of kitchen wastes in two-phase anaerobic digestion. *Environmental technology*, 3, 329-339.

Zuo, Z., Wu, S., Zhang, W., Dong, R. 2014. Performance of two-stage vegetable waste anaerobic digestion depending on varying recirculation rates. *Bioresource Technology*, 162, 266-272.

Chapter 6: General discussion, conclusion and recommendations

6.1 General discussion

Addressing the global sanitation crisis requires effective and sustainable interventions for faecal sludge management. Worldwide, 2.7 billion people are using onsite sanitation systems, with a number expected to increase to 5 billion by 2030 (Strande *et al.*, 2014) due to rapid population growth and increasing emergence of LIHDS in cities of developing countries. Zooming in on Kenya, more than 100 LIHDS in Nairobi have emerged owing to a lag in planning and development of infrastructure to meet the demands of the growing population (AWF, 2013). These settlements, as reported in other countries, are characterised by poor sanitation, haphazard development, high population, high poverty levels and insecure land tenure (Katukiza *et al.*, 2010; Kulabako *et al.*, 2010; Mels *et al.*, 2009; Scott *et al.*, 2013).

Consequences of poor sanitation are dire; owing to the high pathogenic load in excreta, it poses high environmental and public health risks (Winblad & Simpson-Hebert, 2004), (Feachem *et al.*, 1983). Engineers and city planners previously focused on conventional sewer-based sanitation systems, which, in addition to being expensive to develop (Lalander *et al.*, 2013; Mara, 2013), require significant costs in maintaining and upgrading the infrastructure (Kone, 2010; Schertenleib, 2005; Zimmer & Hofwegen, 2006). Onsite sanitation systems, previously viewed as a sanitation option solely viable for rural areas, were adopted as a sanitation solution for the urban LIHDS (Strande *et al.*, 2014). However, thus far, there are no, or very limited, proper management systems available for the faecal sludge (FS) generated by the onsite systems, compromising public health (Mberu *et al.*, 2016). Recent technological advances in sanitation options has resulted in different onsite sanitation technologies, e.g. urine diverting dry toilets (UDDTs), pee poo bags, pour flush toilets connected to septic tanks, etc. (Katukiza *et al.*, 2010). However, thus far, these technologies have not been adopted for widespread usage with about 98% of the sanitary facilities remaining being pit latrines (See Chapter 2, Figure 2.3); consequently, sanitation conditions remain dire (Gulis *et al.*, 2004; Mberu *et al.*, 2016; Zulu *et al.*, 2011). Chapter 2 describes the impediments to LIHDS sanitation enhancement in Kibera, Kenya.

In Kenya, there is no policy framework governing planning, implementation and management of onsite sanitation within the settlements (Mansour *et al.*, 2017). In addition, institutional set-up and regulatory functions of urban sanitation sector are shared by different institutions leading to un-coordinated/unregulated activities among the stakeholders involved along the sanitation chain. For example, whereas sewerage services fall under the authority of the Ministry of Water, onsite sanitation is under the Ministry of Health (See Chapter 2). Moreover, NCWSC is mandated to provide FS disposal points, NCC is mandated to issue business permits for sanitation business and enforce public health rule while NEMA regulates discharge into the sewerage system.

Sanitation improvement efforts among LIHDS, mainly focus on provision of sanitation facilities (human interface), neglecting sludge management and disposal (See Chapter 2, Section 3.1.1). Zooming in on Kenya, Nairobi City Water and Sewerage Company is mandated to provide FS disposal points, either to sewage treatment plants or at designated manholes along the sewer line. This has not been actualised, aggravating FS management, with the demand for service outstripping the available supply. Owing to overcrowding, lack of vehicular access, and lack of potential for investment in the sector, 79% (see chapter 2, Figure 2.5a) of pit emptying services are provided by illegal pit emptiers and 18% by registered manual pit emptiers. Moreover, due to lack of FS disposal points, 85% (See Chapter 2, Figure 2.6) of all sludge is disposed to the environment untreated either directly to river bodies or storm water drain, causing health problems in the communities (Gulis *et al.,* 2004; Mberu *et al.,* 2016; Zulu *et al.,* 2011). Moreover, health and safety concerns were not a priority among the pit emptiers (See Chapter 2, Figure 2.7). Lack of disposal sites and protective gear for, pit emptiers are oblivious of the health risk it poses to the community. Training on safe emptying practices is critical to reverse this trend.

Operation and maintenance of sanitation facilities play a distinct role in their overall sustainability (See Chapter 2, Section3.1.4). Pay and use approach of sanitation provision enhances operation and maintenance initiatives. Whereas in 73% of the free to use community facilities were abandoned on fill up, 89% and 77% (See Chapter 2, Figure 2.4) of community-based organisation (CBOs) and entrepreneur managed facilities, respectively, were well managed. In CBO and entrepreneur managed facilities, a fee is charged per use of the facility, which ranges between (\$ 0.03-0.1) per every use. The fee that is charged is used to facilitate management operations within the facility i.e. cleaning and emptying.

Adoption of pit latrines in Kibera remains high 98% (See Chapter 2, Figure 2.3), despite recent advances in technological development on other onsite sanitation technologies, such as, UDDTs, pee poo bags, pour flush toilets connected to septic tanks etc. (Katukiza *et al.,* 2010). Apparently, these novel technologies remain non-recognised during sanitation planning. Thus far, the sanitation provision approach is supply driven where cost and space are key considerations in identifying the most suitable technology, neglecting accessibility, social-cultural issues and soil conditions (See Chapter 2, Figure 2.2).

Partnership-based sanitation provision improvements provide an entry point for broader initiatives to improve living conditions in informal settlements. With regard to the LIHDS located in or near Nairobi, the Umande Trust business model provides such partnership-based sanitation provision. Umande Trust is a non-governmental organisation (NGO) and is registered by National Environmental Management Authority (NEMA) and Nairobi City Council (NCC). Their business model combines basic service provision with economic empowerment of the communities (http://umande.org/). They enlist services of CBOs in management of bio-centres, who thereafter contract registered pit emptiers to offer emptying services. In Accra, Ghana, a partnership between government and private sector enhanced faecal sludge management (Boot & Scott, 2008). A similar case was reported in Bamako, Mali (Marc *et al.,* 2004).

The need for implementing an effective treatment technology for FS collected from LIHDS can't be underrated. While conducting our a survey carried out in this study (Chapter 2, Section 3.1.4), 88% of facilities visited were operational signifying that emptying services were functional. However, 85% of the collected FS ends untreated in the environment via unhygienic disposal pathways (See Chapter 2, Figure 2.6(b)). Moreover, enterprises that have adopted other sanitation options, such as UDDT, also require treatment options for the collected FS. Treatment of UDDT-FS is emphasised in literature, since addition of ash and sawdust after toilet use is insufficient for pathogen inactivation (Niwagaba *et al.*, 2009a). A case in point is Sanergy- Kenya, a social enterprise working on sanitation enhancement within LIHDS in Kenya. In the business model of Sanergy, FS is collected in specially designed sealable, removable containers (sometimes called cartridges). Serving about 100,000 LIHDS residents per day, it collects and transports approximately 6 Kilo tonnes of UDDT-FS annually (http://www.sanergy.com/). By ensuring that the collected FS is well treated before disposal/ reuse, the health risk associated with open dumping of FS would be reduced (See Chapter 3, Section 3.3).

Owing to lack of space, decentralised treatment units are a viable option for management of FS from LIHDS. In particular, anaerobic digestion (AD) provides an attractive approach in FS treatment (Rajagopal *et al.*, 2013) as it can be implemented at any scale, either onsite or offsite. In addition to FS stabilisation, AD offers biochemical energy recovery through methane production from the biodegradable organic matter in the FS. Moreover, the resulting liquid flows from AD plants are characterised by high concentration of nutrients giving an effluent with good fertilising properties (Avery *et al.*, 2014; Fonoll *et al.*, 2015; Nallathambi Gunaseelan, 1997; Romero-Güiza *et al.*, 2014). However, when applying UDDT-FS as the sole substrate, the application of AD for the treatment of FS has been limited by unsatisfactory pathogen inactivation (Chaggu, 2004; Dudley *et al.*, 1980; Foliguet & Doncoeur, 1972; Leclerc & Brouzes, 1973; McKinney *et al.*, 1958; Pramer *et al.*, 1950) in addition to low methane production (Fagbohungbe *et al.*, 2015; Rajagopal *et al.*, 2013). Microbiological safety of the liquid digestate and/or treated sludge is essential, especially when reuse of the treated matter for agricultural purposes is considered (Avery *et al.*, 2014), as it can lead to transmission of enteric diseases (Pennington, 2001; Smith *et al.*, 2005). These human health risks are also of concern when disposal to the environment is considered. As such, in our study, Chapter 3, 4, and 5 explored enhanced pathogen inactivation and biochemical energy recovery during anaerobic stabilisation of FS.

6.1.1 Enhancing pathogen inactivation during AD

Application of UDDT-FS as the sole substrate for AD is characterised by unsatisfactory pathogen inactivation (Chaggu, 2004; Dudley *et al.*, 1980; Foliguet & Doncoeur, 1972; Leclerc & Brouzes, 1973; McKinney *et al.*, 1958; Pramer *et al.*, 1950). The low pathogen inactivation is associated to the high UDDT-FS buffer capacity during the digestion process (Fonoll *et al.*, 2015; Franke-Whittle *et al.*, 2014; Gallert *et al.*, 1998; Murto *et al.*, 2004), which can be reverted by lowering the substrates' pH and increasing the concentrations of volatile fatty acids (VFAs). The latter condition can be achieved by co-digesting FS with carbohydrate-rich substrates such as organic municipal waste (OMW) fractions, which are associated with rapid hydrolysis (see Chapter 3, Section 3.1). OMW enables enhanced VFA build up in the digestion medium, when used as co-substrate and thus increased the non-dissociated (ND)-VFA concentration, particularly when there is a concomitant pH drop (see also Chapter 4).

The pH and VFA concentrations in the digester both determine the bacterial survival rate during AD (Abdul & Lloyd, 1985; Farrah & Bitton, 1983; Sahlström, 2003). VFA toxicity is associated with the presence of non-dissociated acid molecules: ND-VFAs are able to passively diffuse through the cell membrane of microbes and will dissociate internally, disturbing internal pH, impacting protein's tertiary structure, and inhibiting microbial growth (Jiang et al., 2013; Wang et al., 2014a; Zhang et al., 2005). Additionally, ND-VFAs can make the cell membrane permeable, which allows leaching of the cell content and disintegration of the microbes. The antibacterial effects of ND-VFA have been demonstrated in treatment of enteric E. coli infections of rabbits and pigs, where a rise in caecal pH in diarrhoeic conditions over the normal pH was cited as the main infection cause (Prohászka, 1980b; Prohászka, 1986b): at higher pH, less ND-VFA was present to inactivate pathogens.

The potential for utilising VFAs as a sanitising agent during anaerobic digestion of UDDT-FS was evaluated in Chapter 3. Initial laboratory experiments applied store bought VFAs (acetate, butyrate and propionate) and experimental results indicated increased pathogen inactivation at higher VFA concentrations (Chapter 3). When applying UDDT-FS as the sole substrate, incomplete pathogen inactivation was attained, and was associated with low VFA build-up. On spiking UDDT-FS substrates with store bought VFAs, E. coli inactivation up to 4.7 log units/ day was achieved, as compared to UDDT-FS control sample that only achieved 0.6 log units/ day. Effects of co-digesting UDDT-FS and OMW on VFA build up were evaluated applying different UDDT-FS:OMW mix ratios. By co-digesting human waste (UDDT-FS) with mixed organic market waste (OMW), acid formation is enhanced, since OMW is carbohydrate rich and easily hydrolysable. Increasing the OMW fraction in the feed substrate led to rapid acidification, thereby lowering the pH and increasing the ND-VFA concentration, particularly when a concomitant pH drop was observed. An ND-VFA concentration of 4800-6000 mg/L was required to achieve E. coli log inactivation to below detectable levels and complete A. lumbricoides egg inactivation in less than four days. E. coli and A. lumbricoides egg inactivation was found to be related to the concentration of ND-VFA, increasing with an increase in the OMW fraction in the feed substrate. The increased concentration of ND-VFAs induced a toxic effect not only on pathogens but also on the anaerobic bacterial population, including the VFA producers and methanogenic Archaea, requiring optimisation of OMW dosing. In addition, optimisation of OMW dosing is critical to ensure optimal volumes of UDDT-FS being treated and reduce associated transportation cost for OMW in case of off-site treatment. As such, the UDDT-FS:OMW ratio of 4:1 was recommended for further tests under pilot scale applications. Field experiments were set-up to further evaluate the laboratory-scale results on ND-VFA build-up and pathogen inactivation.

Pilot scale experiments evaluated the performance of single and two stage plug-flow reactors with respect to pathogen inactivation (Chapter 4). Whereas in single stage plug-flow reactors, acidogenesis/hydrolysis and methanogenesis stages occurred in same reactor, the two-stage plug-flow reactors physically separated the two stages. The first stage is characterised by enhanced hydrolysis and acidification of the organic matter, whereas the produced acids are methanised in the second stage. The tests applied UDDT-FS:OMW ratio 4:1, 12% total solids (TS) concentration, with 12% TS being maximum TS concentration that could flow through the plug-flow reactor without any need for mechanical pumping. *E. coli*, one of the indicator organisms for the possible use of digestate coming from faecal matter in agriculture was used as an indicator organism for pathogen inactivation. The two-stage plug-flow reactor showed better pathogen inactivation than the single stage plug-flow reactor. While operating the two-stage plug-flow reactor, caution must be taken to avoid reactor acidification. Highest VFA concentrations of 16.3±1.3 g/l were obtained at a pH of 4.9 in the hydrolysis/acidogenesis first stage reactor, applying a UDDT-FS:OMW ratio of 4:1 and 12% TS, corresponding to a non-dissociated (ND)-VFA concentration of 6.9±2.0 g/l. The corresponding decay rate reached a value of 1.6 /d. This pH value inhibits methanogenic reactions; thus to prevent acidification of the subsequent methanogenic reactor, the low pH effluent with high VFA concentrations requires pH correction for stable methanogenesis. Results showed that pH adjustment achieved via methanogenic effluent recycle was sufficient to maintain efficient methanogenesis in the second stage reactor. A decay rate of 1.1/d was attained within the first third of the methanogenic reactor length, which declined to 0.6/d within the last third part of the reactor length. *E. coli* inactivation to below undetectable levels was achieved, showing the potential of the hygienised effluent for agricultural applications. In the parallel operated sole substrate digestion of UDDT-FS, a 6 and 6.5-fold lower concentration of ND-VFA was observed between the first two sampling points, owing to the prevailing relatively high local pH. Subsequently, with the low achieved ND-VFA build-up, the reactor's effluent pathogen inactivation was low, i.e. only 3.0 *E. coli log* removal. The relatively low pH between the first two sampling points in the single stage co-digestion reactor, which resulted from OMW hydrolysis/acidification as shown by the increasing ND-VFA build-up to 0.6g/l-2.4 g/l, emphasises the importance of a proper UDDT-FS:OMW ratio, avoiding full system acidification and potential failure.

6.1.2 Biochemical energy recovery during anaerobic stabilisation of UDDT-FS

Low methane production is a key drawback during anaerobic digestion of faecal matter owing its high buffer capacity (Rajagopal *et al.* 2013; Fagbohungbe *et al.* 2015). Chapter 5 evaluated the potential biochemical energy recovery during AD of UDDT-FS, both at laboratory scale and pilot scale. Laboratory scale biochemical methane potential (BMP) tests showed increasing methane production with increasing OMW fraction in the feed substrate. The methane production rate during solely UDDT-FS digestion only was 269.4 mL CH_4/g VS added, whereas co-digestion with OMW resulted in a methane production rate of 315.9 mL CH_4/g VS added The highest methane production was observed during digestion of OMW as the sole substrate, reaching a value of 521.0 mLCH_4/g VS added Care had to be taken to prevent reactor acidification. Pilot scale experiments researched the performance of single stage and two stage plug-flow reactors applying UDDT-FS:OMW ratios of 4:1, at 12% TS concentration. Whereas the mix ratio was recommended from laboratory scale BMP tests, the applied TS concentration was chosen as this was the highest TS concentration that still could freely flow without the need for mechanical pumping, as observed in a series of experiments. The lowest methane production was observed during digestion of UDDT-FS as the sole substrate, i.e. 271 ± 13 mL CH_4/ g VS$_{added}$. During co-digestion with OMW a higher methane production was achieved, which was similar for both the single stage reactor ($R_{s-4:1, 12\%}$), i.e. 314 ± 15 mL CH_4 /g VS added, and the two stage reactor $R_{am-4:1, 12\%}$, i.e. 325 ± 12 mL CH_4 /g VS added, applying 12% TS slurry concentration. Remarkably, the biogas production in the two stage plug-flow reactor reached 571 ± 25 mL biogas/g VS added, which was about 12% higher than in the $R_{s-4:1, 12\%}$. This difference was attributed to enhanced waste solubilisation in the acidifying stage and increased CO_2 dissolution, resulting from mixing the bicarbonate-rich methanogenic effluent for neutralisation purposes with the low pH (4.9) influent coming from the pre-acidification stage. The single stage plug-flow reactor as such had a higher quality biogas in terms of methane concentration than the two stage reactor.

6.1.3   Sanitation in relation to energy crisis

Kenya's development plan under Vision 2030 anticipates rapid increase in energy demand arising from economic and social activities that will be undertaken (Mugenya, 2012), despite escalating electricity cost at USD 0.15\$ /kWh (IEA, 2015). The high costs has been associated with increasing oil prices and peaking of hydropower generated electricity due to chronic drought and impact of deforestation on river water supply (Arati, 2006). The key available energy supply options include biomass 74.6%, petroleum (19.1%), electricity generated from hydro-power, geothermal and wind (5.9%) and coal (0.4%) (Karekezi et al., 2008). Owing to effects of the Global Corona Virus (COVID-19) pandemic being experienced in year 2020, a reversal in trend on increasing energy prices in Kenya has been observed https://www.fitchsolutions.com/corporates/retail-consumer/kenya-and-covid-19-impact-consumer-sector-06-05-2020. This was done to lower household expenditure thus cautioning the population against negative COVID-19 impacts.

In urban informal settlements, a direct correlation has been established between poverty levels and type and quantity of energy used, which in turn play a vital role in the services (Karekezi et al., 2008). Also in urban informal settlements activities are carried out that are energy-intensive both at household and small and medium enterprises, such as food kiosks, food vending, welding and carpentry workshops, garages and carwashes (Karekezi et al., 2008). The high energy demand and its associated high cost has led to espousal of crude practices such as illegal electricity connections which result to rampant fire outbreak incidences. Moreover, the wide spread use of charcoal/ wood as fuel source has led to increased deforestation (Kiplagat et al., 2011) and a high rate of acute respiratory illnesses calling for need to venture into other fuel sources (Gulis et al., 2004).

Providing novel forms of energy supply at local level by utilising renewable energy sources, such as biogas, will play a role in addressing the energy crisis. This will be achieved by harvesting the biochemical energy from organic solid waste and utilising it for power generation within the local settlements. This is a cost-effective option in comparison to the centralised electricity grid that is accompanied by high energy costs. Within Kibera LIHDS, Nairobi, the potentials for the application of anaerobic digesters for biochemical energy recovery from faecal sludge has been demonstrated at field scale, with the sanitary bio-centres finding its niche within the settlements as sanitation solution (Umande Trust, 2007). Biogas produced from the facilities is sold to the LIHDS dwellers, as cooking fuel within the centres or to provide hot showers. The charges depend on the type of service offered and range from 0.2 euro (rice cooking charge or hot shower)-0.4 euro (cooking beans). Taking a daily faeces production of 0.5kg/ person, 1.2 million kg faeces/ day can be made available for biogas production within Nairobi LIHDS. It should be noted that 60% of Nairobi's four million inhabitants live in its LIHDS.

6.1.4   LIHDS sanitation; sustainability assessment

Sustainable faecal sludge management interventions should provide for: 1) collection, transportation, treatment and disposal/or reuse, and 2) Ensure continued operations and maintenance of the facilities upon completion. Among the informal settlements, the reliance of aid to finance operations in sanitation service delivery is a critical challenge requiring redress. Such was demonstrated in free community facilities where 73% were abandoned after use. In such facilities, operation and maintenance operations are neglected. Three approaches with potential for enhancing sanitation situations are:

i.   Household based approach- This demonstrated to be the best model for sanitation provision. All household owned sanitation facilities were well managed. However, this is limited by availability of space and cost (Chapter 2).

ii.   Pay and use system-Sanitation facilities applying a pay and use approach demonstrated good operation and maintenance operations: 77% and 89% of entrepreneurship and CBO based sanitation facilities respectively. Operation and maintenance operations are ensured by proceeds from using the facilities.

iii.   Partnership based approach to sanitation provision- The model currently adopted by Umande Trust provides a viable approach for attaining social objective of inclusive sanitation for underprivileged population. They combine basic service provision with economic empowerment of the LIHDS communities. Through donor funding, Umande Trust, under supervision of their technical team, engages trained community technical teams (CTT) to install the sanitary Bio-centres. These facilities serve as multipurpose projects since they provide a range of crucial services to the communities. The management of the Bio-centres is by registered CBOs or self-help groups who through a competitive and transparent process win the process. The identified CBO manages the centre on a pay and use system, whereas the community centre is rented on a monthly basis. A laid-out structure has been provided by Umande Trust that shows how the proceeds from the Bio-centre are used: 60% of the money is saved in CBOs bank which are later divided among the members, 30% of the money caters for operation and maintenance cost and includes cleaning and desludging, whereas 10% is deposited in Umande Trust account and facilitates other related sanitation projects within the settlements (See Chapter 2).

For sustainability, interventions towards sanitation enhancement in these settlements need supportive government policies, which requires creating an enabling environment. This involves facilitating and recognising the interventions as a viable alternative to sewers in informal settlements and hard-to-reach areas. Whereas rich residents of low-income countries receive a subsidised service despite being much more able to pay (Gerlach & Franceys, 2010), the poor pay a considerable proportion of their income for services like pit latrine emptying, which still present considerable health hazards (Thye *et al.*, 2011). Sanitation is a public good, and despite the noble intentions of sanitation interventions, it should not be left to the free market. Policies are needed that recognise sanitation interventions among informal settlements just like the sewer-based sanitation.

6.2   Conclusions

The overall objective of this study was to evaluate impediments to LIHDS sanitation enhancement and to recommend a technology for the management of generated FS. A policy guideline governing implementation and operations of onsite sanitation is critical. This would ensure coordinated activities among various stakeholders involved along the sanitation chain. The study findings have shown a high potential for applying anaerobic digestion as a treatment option for UDDT-FS. The following aspects should however be considered:

a. Substrate type- The research findings recommend co-digestion of UDDT-FS and OMW. When using FS as a sole digestion substrate, low pathogen inactivation was achieved, in addition to low methane production. Increased pathogen inactivation was observed at an increasing OMW fraction in feed substrate. However, the application of high OMW fractions is disadvantageous due to: 1) Logistic concerns due to collection, sorting and transportation costs of the waste from the LIHDS to the treatment site, 2) At high OMW fraction e.g. UDDT-FS:OMW ratios 1:4, 1:2 and 1:1, the pH declines to very low levels, which may be toxic to *E. coli,* as well as all other microbial populations. For practical purposes ND-VFA concentration of approximately 2800-4300 mg/L, is sufficient causing between 3-5 *E. coli* log inactivation in four days. This is achieved by co-digestion of UDDT-FS:OMW mix ration 4:1, at 35^0C.

b. Total solids concentration- Operating single stage and two stage plug-flow reactors at a UDDT-FS:OMW ratio of 4:1, at 12% TS concentration, maximised the UDDT-FS fraction that could be treated per unit time. A TS concentration of 12% was the highest possible concentration that could flow without the need for mechanical pumping. In addition, a higher pathogen inactivation and methane production rate was achieved at 12% TS concentration compared to 10% TS concentration.

c. Reactor configuration- The two stage plug-flow reactors showed higher pathogen inactivation than corresponding single stage reactors, at both 12% and 10% TS concentration. Although both reactors showed comparable methane production, better quality biogas was observed in the two-stage reactor.

6.3   Recommendations

Recommendations of this study have been classified into three categories;

a. Technical performance

The acidogenic/ hydrolysis reactor and methanogenic reactor were separate entities, where the pre-acidified effluent from the first reactor, after pH adjustment, was used as feed for the methanogenic reactor. A re-design of the setup to provide for direct flow of the pre-acidified effluent to the methanogenic reactor would aid in maintaining anaerobic conditions inside the reactors, and likely a higher biogas production. The proposed design is shown in Figure 6.1.

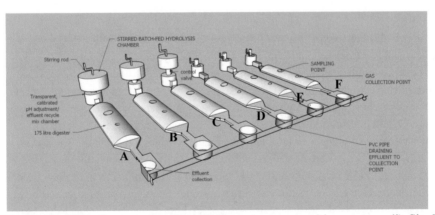

Figure 6.1: Plug flow reactor; i) A-C-Two stage anaerobic reactors, ii) Single stage anaerobic reactors

b. Technical full-scale application

Owing to the better FS pathogen removal, the two-stage plug-flow reactor setup (Figure 6.1) is recommended for full-scale application.

The system has potential for application as an off-site FS treatment technology at any scale, receiving any type of faecal matter, collected from different types of sanitary systems, e.g.

UDDTq, peepoo bags etc. To reduce logistics and operation cost of transporting UDDT-FS, off site treatment sites should be located as near as possible to the UDDT-FS collection points. On-site applications will reduce transportation costs to a minimum and are certainly worthwhile to be further studied. Moreover, on-site arrangements may entail connections of pour flush/ cistern flush toilets directly to the anaerobic digester; a solution certainly of interest in the low income – high density settlements.

      c. The performed research recommends formulation and implementation of a
         Government policy document that recognises non-sewered sanitation as an alternative
         to sewer-based sanitation.

References

Arati, J. 2006. Evaluating the economic feasibility of anaerobic digestion of Kawangware market waste, MSc Thesis, University of Wisconsin- River Falls, 2006.

Avery, L.M., Anchang, K.Y., Tumwesige, V., Strachan, N., Goude, P.J. 2014. Potential for Pathogen reduction in anaerobic digestion and biogas generation in Sub-Saharan Africa. *Biomass and Bioenergy*, 70(0), 112-124.

AWF. 2013. Expanding branded toilet entrepreneurship for improved sustainable sanitation in poor neighbourhoods of Nairobi, Kenya.

Boot, N., Scott, R. 2008. Facal sludge management in Accra, Ghana: strengthening links in the chain: 33rd WEDC International Conference, Accra, Ghana, 2008. http://wedc.lboro.ac.uk/resources/conference/33/Boot_NLD.pdf.

Chaggu, E.J. 2004. Sustainable Environmental Protection Using Modified Pit-Latrines. Ph.D Thesis, Wageningen University, The Netherlands.

Dudley, D.J., Guentzel, M.N., Ibarra, M.J., Moore, B.E., Sagik, B.P. 1980. Enumeration of potentially pathogenic bacteria from sewage sludge. *Applied Environmental Microbiology*, 39, 118-126.

Fagbohungbe, M.O., Herbert, B.M.J., Li, H., Ricketts, L., Semple, K.T. 2015. The effect of substrate to inoculum ratios on the anaerobic digestion of human faecal material. *Environmental Technology & Innovation*, 3(0), 121-129.

Feachem, R.G., Bradley, D.J., Garelick, H., D., M.D. 1983. Sanitation and Disease Health Aspects of Excreta and Wastewater Management. *Report No.:11616 Type: (PUB)*

Foliguet, J.M., Doncoeur, F. 1972. Inactivation in fresh and digested wastewater sludges by pasteurization. *Water Research*, 6, 1399-1407.

Fonoll, X., Astals, S., Dosta, J., Mata-Alvarez, J. 2015. Anaerobic co-digestion of sewage sludge and fruit wastes: Evaluation of the transitory states when the co-substrate is changed. *Chemical Engineering Journal*, 262(0), 1268-1274.

Franke-Whittle, I.H., Walter, A., Ebner, C., Insam, H. 2014. Investigation into the effect of high concentrations of volatile fatty acids in anaerobic digestion on methanogenic communities. *Waste Management*, 34(11), 2080-2089.

Gallert, C., Bauer, S., Winter, J. 1998. Effect of ammonia on the anaerobic degradation of protein by a mesophilic and thermophilic biowaste population. . *Applied Microbiology and Biotechnology* (50), 495-501.

Gerlach, E., Franceys, R. 2010. Regulating Water Services for All in Developing Economies. *World Development*, 38, 1229-1240.

Gulis, G., Mulumba, J.A.A., Juma, O., Kakosova, B. 2004. Health status of people of slums in Nairobi, Kenya. *Environmental Research*, 96(2), 219-227.

IEA. 2015. Situational Analysis of Energy Industry, Policy and Strategy for Kenya.

Jiang, J., Zhang, Y., Li, K., Wang, Q., Gong, C., Li, M. 2013. Volatile fatty acids production from food waste: Effects of pH, temperature, and organic loading rate. *Bioresource Technology*, 143(0), 525-530.

Karekezi, S., Kimani, J., Onguru, O. 2008. Energy access among the urban poor in Kenya. *Energy for Sustainable Development*, 12(4), 38-48.

Katukiza, A.Y., Ronteltap, M., Oleja, A., Niwagaba, C.B., Kansiime, F., Lens, P.N.L. 2010. Selection of sustainable sanitation technologies for urban slums — A case of Bwaise III in Kampala, Uganda. *Science of The Total Environment*, 409(1), 52-62.

Kiplagat, J.K., Wang, R.Z., Li, T.X. 2011. Renewable energy in Kenya: Resource potential and status of exploitation. *Renewable and Sustainable Energy Reviews*, 15(6), 2960-2973.

Kone, D. 2010. Making urban excreta and wastewater management contribute to cities economic development: a pradigm shift *Water Policy*, 12, 602-610.

Kulabako, N., Nalubega, M., Wozei, E., Thunvik, R. 2010. Environmental health practices, constraints and possible interventions in peri-urban settlements in developing countries- a review of Kampala, Uganda. *Int J Environ Health Res* 20(4), 231-257.

Lalander, C.H., Hill, G.B., Vinnerås, B. 2013. Hygienic quality of faeces treated in urine diverting vermicomposting toilets. *Waste Management*, 33(11), 2204-2210.

Leclerc, H., Brouzes, P. 1973. Sanitary aspects of sludge treatment. *Water Research*, 7(3), 355-360. Mansour, G., Oyaya, C., Owor, M. 2017. Water and for the urban poor: Sanitation and Situation analysis of the urban sanitation sector in Kenya.

Mara, D. 2013. Pits, pipes, ponds – And me. *Water Research*, 47(7), 2105-2117.

Marc, J., Doulaye, K., Martin, S. 2004. Private sector management of faecal sludge: A model for the future? Focus on an innovative planning experience in Bamako, Mali. Department of Water and Sanitation in Developing Countries (SANDEC), Switzerland.

Mberu, B.U., Haregu, T.N., Kyobutungi, C., Ezeh, A.C. 2016. Health and health-related indicators in slum, rural, and urban communities: a comparative analysis. African Population and Health Research Center, Nairobi, Kenya.

McKinney, R.E., Langley, H.E., Tomlinson, H.D. 1958. Survival of Salmonella typhosa during anaerobic digestion. I. Experimental methods and high rate digester studies. *Sewage Ind. Wastes*, 30, 1467-1477.

Mels, A., Castellano, D., Braadbaart, O., Veenstra, S., Dijkstra, I., Meulman, B., Singels, A., Wilsenach, J.A. 2009. Sanitation services for the informal settlements of Cape Town, South Africa. *Desalination*, 248(1–3), 330-337.

Mugenya, P. 2012. Energy from biomass & biogas : Market opportunities in kenya; Opportunities for bioenergy solutions in the Kenyan sugar industry.

Murto, M., Bjo¨rnsson, L., Mattiasson, B. 2004. Impact of food industrial waste on anaerobic co-digestion of sewage sludge and pig manure. *Journal of Environmental Management* 70, 101-107.

Nallathambi Gunaseelan, V. 1997. Anaerobic digestion of biomass for methane production: A review. *Biomass and Bioenergy*, 13(1–2), 83-114.

Niwagaba, C., Kulabako, R.N., Mugala, P., Jönsson, H. 2009. Comparing microbial die-off in separately collected faeces with ash and sawdust additives. *Waste Management*, 29(7), 2214-2219.

Pennington, T.H. 2001. Pathogens in agriculture and the environment. In: Pathogens in Agriculture and the Environment, Meeting organised by the SCI Agriculture and Environment Group, 16 October, SCI, London.

Pramer, D., H. , Heukelekian, Ragotskie, R.A. 1950. Survival of tubercule bacilli in various sewage treatment processes. I. Development of a method for the quantitative recovery of mycobacteria from sewage. *Public Health Rep.* , 65, 851-859.

Prohászka, L. 1980. Antibacterial Effect of Volatile Fatty Acids in Enteric *E. coli*-infections of Rabbits. Zentralblatt für Veterinärmedizin Reihe B, 27(8), 631-639.

Prohászka, L. 1986. Antibacterial Mechanism of Volatile Fatty Acids in the Intestinal Tract of Pigs agains Escherichia coli. Journal of Veterinary Medicine, Series B, 33(1-10), 166-173.

Rajagopal, R., Lim, J.W., Mao, Y., Chen, C.-L., Wang, J.-Y. 2013. Anaerobic co-digestion of source segregated brown water (feces-without-urine) and food waste: For Singapore context. *Science of The Total Environment*, 443(0), 877-886.

Romero-Güiza, M.S., Astals, S., Chimenos, J.M., Martínez, M., Mata-Alvarez, J. 2014. Improving anaerobic digestion of pig manure by adding in the same reactor a stabilizing agent formulated with low-grade magnesium oxide. *Biomass and Bioenergy*, 67, 243-251.

Schertenleib, R. 2005. From conventional to advanced environmental sanitation. *Water Science & Technology*, 51(10), 7-14.

Scott, P., Cotton, A., Sohail Khan, M. 2013. Tenure security and household investment decisions for urban sanitation: The case of Dakar, Senegal. *Habitat International*, 40(0), 58-64.

Smith, S.R., Lang, N.L., Cheung, K.H.M., Spanoudaki, K. 2005. Factors controlling pathogen destruction during anaerobic digestion of biowastes. *Waste Management*, 25(4), 417-425.

Strande, L., Ronteltap, M., Brdjanovic, D. 2014. Faecal Sludge Management Systems Approach for Implementation and Operation. IWA publishing, ww.iwapublishing.com.

Thye, Y.P., Templeton, M., Ali, M. 2011. A Critical Review of Technologies for Pit Latrine Emptying in Developing Countries. *Critical Reviews in Environmental Science and Technology*, 41, 1793-1819.

Wang, K., Yin, J., Shen, D., Li, N. 2014. Anaerobic digestion of food waste for volatile fatty acids (VFAs) production with different types of inoculum: Effect of pH. *Bioresource Technology*, 161(0), 395-401.

Winblad, U., Simpson-Hebert, M. 2004. *Ecological sanitation - revised and enlarged edition.* Stockholm Institute of Environment, Stockholm, Sweden.

Zhang, B., Zhang, L.L., Zhang, S.C., Shi, H.Z., Cai, W.M. 2005. The influence of pH on hydrolysis and acidogenesis of kitchen wastes in two-phase anaerobic digestion. *Environmental technology*, 3, 329-339.

Zimmer, D., Hofwegen, P. 2006. Costing MDG Target 10 on water supply and sanitation: comparative analysis, obstacles and recommendations.

Zulu, E.M., Beguy, D., Ezeh, A., C., Bocquier, P., Madise, N.J., Cleland, J., Falkingham, J. 2011. Overview of migration, poverty and health dynamics in Nairobi City's slum settlements. *Journal of Urban Health: Bulletin of the New York Academy of Medicine, Vol. 88, Suppl. 2 doi:10.1007/s11524-011-9595-0.*

Netherlands Research School for the
Socio-Economic and Natural Sciences of the Environment

DIPLOMA

for specialised PhD training

The Netherlands research school for the
Socio-Economic and Natural Sciences of the Environment
(SENSE) declares that

Joy Nyawira Riungu

born on 3 December 1974 in Meru, Kenya

has successfully fulfilled all requirements of the
educational PhD programme of SENSE.

Delft, 28 January 2021

Chair of the SENSE board

Prof. dr. Martin Wassen

The SENSE Director

Prof. Philipp Pattberg

The SENSE Research School has been accredited by the Royal Netherlands Academy of Arts and Sciences
(KNAW)

K O N I N K L I J K E   N E D E R L A N D S E
A K A D E M I E   V A N   W E T E N S C H A P P E N

The SENSE Research School declares that Joy Nyawira Riungu has successfully fulfilled all requirements of the educational PhD programme of SENSE with a work load of 53.3 EC, including the following activities:

SENSE PhD Courses

o   Environmental research in context (2013)
o   Research in context activity: 'Creating and disseminating of instructive video presentation on: Resource Oriented Decentralized Sanitation' (2016)

Other PhD and Advanced MSc Courses

o   Sustainable sanitation-decentralized, natural and ecological waste water treatment, Norwegian Institute of life Sciences (2010)
o   Costing sustainable services in sanitation, International Water and Sanitation Centre (2013)
o   Workshop on Maintenance of infrastructure for developing countries, International Science, Technology and Innovation Centre for South-South cooperation (2018)
o   GIS and remote sensing, Jomo Kenyatta University of Agriculture and Technology, (2018)
o   Sanitation systems and services, Meru University of Science and Technology (2020)
o   Leaders in Innovation training (2019)

External training at a foreign research institute

o   Trained on analysis, identification, quantification and inactivation of Helminth (Ascaris) Eggs, Asian Institute of Technology, Thailand (2014)

Management and Didactic Skills Training

o   Founder Sanitation Research Centre, MUST (2019)
o   Supervising MSc student with thesis entitled 'Enhancing sludge hygienisation in anaerobic digestion of UDDT faeces' (2013-2014)

Selection of Oral Presentations

o   *Assessing the bio-methane potential (BMP) and concomitant pathogen removal of UDDT faeces with(out) additional organic substrates.* 3rd Faecal sludge management conference, 18-21 January 2015, Hanoi, Vietnam.
o   *Hygienisation of human waste in anaerobic treatment – making use of intrinsic processes.* IWA Development Congress, 13-16 November 2017, Buenos Aires, Argentina.
o   Simultaneous stabilization, methanation, and hygienization of faecal matter from poor urban settlements applying co-digestion in plug-flow digester systems. *16th IWA World Conference on Anaerobic on digestion, 23-27th June 2019, Delft, The Netherlands.*
o   *Rethinking human waste management,* Sankalp African Summit, 27-28th February 2020 Nairobi, Kenya.

SENSE coordinator PhD education

Dr. ir. Peter Vermeulen